T0212899

Lecture Notes in Computer Science 14149

Founding Editors

Gerhard Goos
Juris Hartmanis

The series Lecture Notes in Computer Science (LNCS), including its subseries Lecture Notes in Artificial Intelligence (LNAI) and Lecture Notes in Bioinformatics (LNBI), has established itself as a medium for the publication of new developments in computer science and information technology research, teaching, and education.

LNCS enjoys close cooperation with the computer science R & D community, the series counts many renowned academics among its volume editors and paper authors, and collaborates with prestigious societies. Its mission is to serve this international community by providing an invaluable service, mainly focused on the publication of conference and workshop proceedings and postproceedings. LNCS commenced publication in 1973.

Andrea Kö · Enrico Francesconi · Adeleh Asemi ·
Gabriele Kotsis · A Min Tjoa · Ismail Khalil
Editors

Electronic Government and the Information Systems Perspective

12th International Conference, EGOVIS 2023
Penang, Malaysia, August 28–30, 2023
Proceedings

 Springer

Editors
Andrea Kö 🅾
Corvinus University of Budapest
Budapest, Hungary

Enrico Francesconi
Italian National Research Council
Florence, Italy

Adeleh Asemi
University of Malaya
Kuala Lumpur, Malaysia

Gabriele Kotsis
Johannes Kepler University Linz
Linz, Austria

A Min Tjoa 🅾
Vienna University of Technology
Vienna, Austria

Ismail Khalil
Johannes Kepler University Linz
Linz, Austria

ISSN 0302-9743   ISSN 1611-3349 (electronic)
Lecture Notes in Computer Science
ISBN 978-3-031-39840-7   ISBN 978-3-031-39841-4 (eBook)
https://doi.org/10.1007/978-3-031-39841-4

This Springer imprint is published by the registered company Springer Nature Switzerland AG
The registered company address is: Gewerbestrasse 11, 6330 Cham, Switzerland

Preface

It is our distinct honor to present the proceedings of the 12th International Conference on Electronic Government and the Information Systems Perspective (EGOVIS 2023), with the theme of Digitalization, Innovation, Automation, and Transformation. This conference served as a platform for experts, scholars, and practitioners to come together and explore the multifaceted landscape of electronic government, delving into the intricate interplay between information systems, digitalization, innovation, automation, and transformation.

In an era where technology is reshaping our societies, electronic government has emerged as a transformative force, revolutionizing the way governments interact with citizens, businesses, and other stakeholders. The seamless integration of information systems, coupled with digitalization efforts, has paved the way for efficient, transparent, and citizen-centric governance models that foster collaboration, improve service delivery, and drive societal progress.

The papers featured in this volume embody the collective efforts of researchers and professionals who are at the forefront of shaping the electronic government landscape. These contributions encompass a wide range of topics, including but not limited to e-governance frameworks, digital service delivery, data analytics, cybersecurity, policy formulation, citizen engagement, and the transformative potential of emerging technologies such as blockchain, artificial intelligence, and Internet of Things.

We are proud to report authors from 10 different countries submitted papers to EGOVIS this year. Our program committee conducted close to 60 reviews, an average of 3 single-blind reviews per submission. From 17 submitted papers the program committee decided to accept 8 full papers with an acceptance rate of 47%.

Each paper included in this compilation has undergone a rigorous review process, ensuring the highest standards of quality and relevance to the conference theme. The depth and diversity of the research presented here highlight the ongoing efforts to leverage information systems and innovative technologies for effective governance, public service provision, and societal transformation.

We would like to express our deepest gratitude to all the authors who have contributed their valuable research and insights to this volume. Their dedication and expertise have enriched the discourse surrounding electronic government, paving the way for novel approaches, best practices, and lessons learned. We also extend our sincere appreciation to the members of the program committee for their meticulous evaluation and valuable feedback that shaped the selection and refinement of the papers.

Furthermore, we want to extend our gratitude to the conference organizers, keynote speakers, and attendees for their unwavering support and enthusiasm. It is through their collective efforts that this conference has become a thriving forum for exchanging knowledge, fostering collaboration, and catalysing meaningful change in the realm of electronic government.

As you delve into the pages of these proceedings, we hope you find inspiration in the insightful research, innovative solutions, and transformative visions shared within. Whether you are a researcher, a practitioner, a policymaker, or an advocate for citizen-centric governance, we believe that the wealth of knowledge presented here will inspire you to explore new horizons and contribute to the ongoing digital transformation of our governments and societies.

Finally, we would like to express our heartfelt appreciation to the entire conference community. It is your dedication, passion, and commitment to advancing electronic government and embracing the information systems perspective that has made this event a resounding success.

August 2023

Andrea Kő
Enrico Francesconi
Adeleh Asemi
Gabriele Kotsis
A Min Tjoa
Ismail Khalil

Organization

Program Committee Chairs

Andrea Kő · Corvinus University of Budapest, Hungary
Enrico Francesconi · Italian National Research Council, Italy and European Parliament, Luxembourg
Adeleh Asemi · University of Malaya, Malaysia

Steering Committee

Gabriele Kotsis · Johannes Kepler University Linz, Austria
A Min Tjoa · Vienna University of Technology, Austria
Robert Wille · Software Competence Center Hagenberg, Austria
Bernhard Moser · Software Competence Center Hagenberg, Austria
Ismail Khalil · Johannes Kepler University Linz, Austria

Program Committee Members

Luis Alvarez Sabucedo · Universidade de Vigo, Spain
Alejandra Cechich · Universidad del Comahue, Argentina
Wichian Chutimaskul · King Mongkut's University of Technology Thonburi, Thailand
Flavio Corradini · University of Camerino, Italy
Peter Cruickshank · Edinburgh Napier University, UK
Vytautas Čyras · Vilnius University, Lithuania
Dirk Draheim · Tallinn University of Technology, Estonia
György Drótos · Corvinus University of Budapest, Hungary
Jean Vincent Fonou-Dombeu · University of KwaZulu-Natal, South Africa
Miguel Ángel Latre · University of Zaragoza, Spain
Fernando Galindo · University of Zaragoza, Spain
Francisco Javier García Marco · Universidad de Zaragoza, Spain
Stefanos Gritzalis · University of Piraeus, Greece
Christos Kalloniatis · University of the Aegean, Greece
Nikos Karacapilidis · University of Patras, Greece
Evangelia Kavakli · University of the Aegean, Greece
Ivan Futo · Multilogic Ltd., Hungary

Christine Leitner	Centre for Economics and Public Administration, UK
Herbert Leitold	A-SIT, Austria
Mara Nikolaidou	Harokopio University, Greece
Javier Nogueras Iso	Universidad de Zaragoza, Spain
Monica Palmirani	Università degli Studi di Bologna, Italy
Aljosa Pasic	Atos Origin, Spain
Aires Rover	Federal University of Santa Catarina, Brazil
A Min Tjoa	Vienna University of Technology, Austria
Costas Vassilakis	University of the Peloponnese, Greece
Chien-Chih Yu	National Cheng Chi University, Taiwan

Organizers

Contents

e-Government

NSFASonto: An Ontology for South African National Student Financial Aid Scheme Operations

Shaneer Luchman, Daniel Mundell, Akaylan Perumal,
and Jean Vincent Fonou-Dombeu(✉) (iD)

School of Mathematics, Statistics and Computer Science,
University of KwaZulu-Natal, Pietermaritzburg, South Africa
{216003502,220108104,216035695}@stu.ukzn.ac.za, fonoudombeu@gmail.com
https://www.ukzn.ac.za/

Abstract. The South African government established the National Student Financial Aid Scheme (NSFAS) to help underprivileged families with finances to access higher education. NSFAS and partner institutions run heterogeneous systems and use different processes and data representations for students and funding operations; which create duplications and inconsistencies in the data, resulting in errors and delays in the approval of students' applications and disbursement of funds. To address these challenges, this study developed an ontology of NSFAS operations, namely, NSFASonto. The resulting NSFASonto ontology was evaluated for reasoning and consistency with the HermiT reasoner and SPARQL queries within Protégé and displayed promising results. The NSFASonto ontology can be leveraged in web-based applications to integrate the NSFAS operations with partner institutions for efficient approval and delivery of financial aid to students across South Africa. Furthermore, the NSFASonto ontology could be reused in other countries to build automated systems for student financial aid operations/processes.

Keywords: Student Financial Aid · Higher Education · NSFAS · Semantic Web · Protégé

1 Introduction

Most universities in South Africa are public universities and are funded by government. However, they still require students to pay tuition fees [1]. A first year student pays on average R64 200 (approximately 4 019 US dollars) per year in a South African university [2]. The problem is that most families in South Africa cannot afford such financial commitments [3]. Therefore, student funding has always been an important issue for the government of South Africa. In 1996, the South African government formed a body called NSFAS to provide funding for students whose families would struggle to pay their tuition fees. NSFAS was given the task of providing loans or bursaries to students whose family household collectively earned less than R350 000 (approximately 21 907 US dollars) annually. In recent years [4], NSFAS was further tasked with providing funds for other student needs such as accommodation, transportation, and textbooks.

Practically, NSFAS works with partner institutions including the Department of Higher Education and Training (DHET), Public Universities (PU), and Technical and Vocational Education and Training (TVET) colleges, to deliver funding to students. This requires the exchange of information amongst NSFAS and its partners. For instance, PU and TVET colleges provide NSFAS with information on students' registration and courses. In turn, NSFAS provides the PU and TVET colleges with funds for students' tuition fees and allowances. However, at present NSFAS and its partner institutions run heterogeneous systems that are not integrated, to process students' finding. The lack of integration between NSFAS system and that of PU and TVET colleges means that each entity has its own representation of students' data; this results in many duplications and multiple storage of same data, which in turn may result in errors and inconsistencies in the information being processed in NSFAS operations. This has created many issues such as (1) delays in the disbursement of allowances to students, (2) inconsistency in the payment patterns by PU and TVET colleges, and (3) communication breakdown between NSFAS and students on the one hand and NSFAS and partner institutions on the other hand [5]. Furthermore, the lack of information has been reported amongst the population about student financial aids worldwide [6] and NSFAS in particular [7,8]; which limits the capability of funding agencies like NSFAS to reach all those in need of funds to pursue higher education.

Semantic web technologies with ontology can be leveraged to address the integration problem amongst NSFAS and its partner institutions as well as to provide a standard common sharable and understandable representation of information of NSFAS financial aid to all South African citizens. An ontology is formal representation of a knowledge domain, through its entities, concepts, objects and the relationships between them [9]. Ontology represents common and shared terminologies of a domain in both conceptual and machine-readable forms, thereby, provides a common understanding of the structure of information to the community of users/actors as well as enables automated processing of knowledge in the domain. Ontology can be used to integrate heterogeneous systems and facilitate their interoperability [10,11]. In fact, the systems that use the same ontology, process the same terminologies, which facilitate their integration and interoperability. Alternatively, if separate ontology has been developed for each system, the terminologies in the ontologies of these systems can be matched to integrate them [11]. The Semantic Web is an extension of the current web in which contents are represented based on their meanings rather than hyperlinks as on the current internet, for both human and machine understanding [10]. The meaning of web content on the Semantic Web is represented with ontology in logic-based languages such as Resource Description Framework Schema (RDFS) and Web Ontology Language (OWL); this enable the development of intelligent web-based applications that can automatically query and reason the web contents to infer new information on behalf of users [10,11]. RDFS and OWL are two standard languages for representing ontologies; they provide features that enable to represent ontologies in a machine-readable form that can be automatically queried and reasoned to infer new knowledge over the web [12]. However, no previous study has attempted to develop ontology for financial aid systems.

This study developed an ontology of NSFAS operations, namely, NSFASonto. Data was collected from various sources including websites, published articles and reports as well as related ontologies. The collected data was used to formalize the NSFASonto ontology in Description Logics (DLs). Description logics is a family of languages that provide syntaxes for logically representing ontology elements such as classes, properties and instances as well as the semantic relations between them. The DLs representation of NSFASonto ontology was implemented in a machine readable form in OWL language within Protégé editor. The resulting NSFASonto ontology was evaluated for reasoning and consistency with the HermiT reasoner and SPARQL queries and displayed promising results. The NSFASonto ontology is a formal data model that can be leveraged in web-based applications to integrate the NSFAS operations with PU and TVET colleges for efficient approval and delivery of funding to students across South Africa. Furthermore, the NSFASonto ontology could be reused in other countries to build automated systems for financial aid operations.

The rest of the paper is organized as follows. The literature is reviewed in Sect. 2. In Sect. 3, the material and methods used to conduct the study are presented and applied to build the NSFASonto ontology. The experimental results are described and explained in Sect. 4. The last section concludes the paper.

1.1   Related Works

Many authors have addressed the topic of student financial aid in higher education [7, 8, 13–17]. The authors [17] carried out a detailed review of literature on student financial aid. The study [13], focused on the reasons for the increased scarcity of financial aid in support of students access to higher education in Sub-Saharan Africa. The authors [14], used empirical data to analyse the impact of NSFAS financial aid on higher education enrolment in South Africa. Another study [8] investigated the experience of first-year students with the NSFAS loan and reported the incapacity of NSFAS to satisfy the high demand for financial support as well as the lack of information about the financial scheme for this category of students. The authors [7], investigated the effect of tuition fees and student loans on the access and success at South African Universities as well as the barriers that hinder the access of underprivileged South Africans to student loans. Another study [15] investigated the reasons why the increase in financial support for access to higher education in the USA did not address the issue of equal choice of colleges between poor, middle class and richest students and proposed solutions to the problem. A similar study [6] reviewed the implementation of student financial aid programs at the federal and state levels in the USA and identified the best practices that contribute to students access and success. A recent study [16], investigated the experiences of the so-called "first-generation college students" on the financial aid process in the USA. The author revealed that this category of students struggles with the completion of the financial aid process.

In light of the above, the current literature focuses on the social and economical aspects of student financial aid and does not address the issues pertaining to the design and adoption of new technologies in support of an increase awareness and access to information as well as integrated online application processes. To the best of our knowledge, no previous study has formally modelled the student financial aid domain to enable machine understanding for automated processing of financial aid operations.

However, authors have developed related ontologies to model higher education institutions [18], student's profile [19], and curriculum [20] in higher education. This study is the first to develop an ontology for the student financial aid domain in higher education. Some concepts such as University, College, School, Department, Module, Course, etc. were gathered from the ontologies reported in the above studies and adopted in the design of NSFASonto ontology.

2   Materials and Methods

2.1   Methodology

The methodology used to develop the NSFASonto ontology is named Ontology Development Methodology 101 (OD101). The OD101 methodology was authored by Noy and McGuinness [21] and has been successfully applied to build ontologies [22,23]. The OD101 methodology is suitable for a rapid development of ontology because it allows one to quickly formalize a knowledge domain in logic-based syntaxes for implementation. Ontology development with the OD101 methodology is done in five main steps including (1) definition of the scope of ontology, (2) reuse of existing ontologies, (3) creating a glossary of terms, (4) definition of classes, properties and instances of the ontology and (5) modeling the target ontology in a logic-based syntax. The three first steps of the OD101 methodology enable to build the vocabulary of the target ontology. This vocabulary then serves in steps 4 and 5 of the methodology. The OD101 methodology is applied in the next subsection to build the NSFASonto ontology.

2.2   Building the Vocabulary of the NSFAS Domain

Various sources of information including website[1], research articles and reports [5,7,8] as well as existing related ontologies [18–20] were investigated to construct the vocabulary of the NSFAS knowledge domain. Competency questions (CQs) were also formulated to investigate NSFAS operations. CQs are questions that the ontology should be able to answer to fulfill its purposes. Examples of CQs formulated include: how can one find information about NSFAS? what information is required to apply for NSFAS? which are the NSFAS approved Universities? which are the NSFAS approved TVET colleges? etc. The collected

---

[1] http://www.nsfas.org.za/content/.

Table 1. Part of FOL Representation of Axioms of the NSFASonto Ontology.

1	$\forall x(Student(x) \rightarrow \exists y(Person(x) \land Institution(y) \land studiesAt(x,y))$
2	$\forall x(SouthAfricanStudent(x) \rightarrow (Student(x) \land SouthAfrican(x)))$
3	$\forall x(EligibleStudent(x) \rightarrow \exists y(SouthAfricanStudent(x) \land$ $\neg hasPreviousQualification(x) \land annualHouseholdIncome(y) \land$ $has(x,y) \land (y < 350000))$
4	$\forall x(EligibleStudent(x) \rightarrow \exists y \exists z(SouthAfricanStudent(x) \land$ $\neg hasPreviousQualification(x) \land annualHouseholdIncome(y) \land$ $Disability(z) \land has(x,y) \land (y < 600000) \land has(x,z))$
5	$\forall x(EligibleStudent(x) \rightarrow \exists y \exists z(SouthAfricanStudent(x) \land$ $\neg hasPreviousQualification(x) \land annualHouseholdIncome(y) \land Year(z) \land$ $has(x,y) \land (y < 122000) \land startedStudying(x,z) \land (z < 2018)$
6	$\forall x(EligibleStudent(x) \rightarrow (SouthAfricanStudent(x) \land$ $\neg hasPreviousQualification(x) \land SASSAGrantRecipient(x))$
7	$\forall x(ApprovedStudent(x) \rightarrow (EligibleStudent(x) \land$ $(\exists y(ApprovedApplication(y) \land submits(x,y))))$
21	$\forall x(Institution(x) \equiv (University(x) \lor TVET(x)))$
28	$\forall x(ApprovedApplication(x) \rightarrow \exists y NSFASEmployee(y) \land isApprovedBy(x,y))$
29	$\forall x(DeniedApplication(x) \rightarrow \exists y NSFASEmployee(y) \land isDeniedBy(x,y))$
39	$\forall x(FundSource(x) \equiv (Donations(x) \lor ExistingFunds(x) \lor$ $GovernmentDepartment(x) \lor PrivateSector(x) \lor$ $LoanRepayment(x) \lor SETA(x)))$
40	$\exists x RecoverySystem(x) \land possesses(NSFAS, x)$
41	$\forall x(RecoverySystem(x) \rightarrow (CollectionAgency(x) \lor Marketing(x) \lor$ $PayrollDeductionAgreement(x)))$
42	$\forall x \exists y(Student(x) \land University(y) \land studiesAt(x,y)) \rightarrow$ $\exists z(TransportationCost(z) \land (requiresFundsFor(x,z) \land (z \leq 7500))))$
43	$\forall x \exists y(Student(x) \land University(y) \land studiesAt(x,y)) \rightarrow$ $\exists z(TextbookCost(z) \land (requiresFundsFor(x,z) \land (z \leq 5200))))$
44	$\forall x \exists y(Student(x) \land University(y) \land studiesAt(x,y)) \rightarrow$ $\exists z(LivingCost(z) \land (requiresFundsFor(x,z) \land (z \leq 15000))))$
45	$\forall x \exists y(Student(x) \land TVET(y) \land studiesAt(x,y)) \rightarrow$ $\exists z(TransportationCost(z) \land (requiresFundsFor(x,z) \land (z \leq 7000))))$
46	$\forall x \exists y(Student(x) \land TVET(y) \land studiesAt(x,y) \land livingArea(x, Urban)) \rightarrow$ $\exists z(AccommodationCost(z) \land (requiresFundsFor(x,z) \land (z \leq 24000))))$
47	$\forall x \exists y(Student(x) \land TVET(y) \land studiesAt(x,y) \land livingArea(x, Peri - Urban)) \rightarrow$ $\exists z(AccommodationCost(z) \land (requiresFundsFor(x,z) \land (z \leq 18900))))$
46	$\forall x \exists y(Student(x) \land TVET(y) \land studiesAt(x,y) \land livingArea(x, Rural)) \rightarrow$ $\exists z(AccommodationCost(z) \land (requiresFundsFor(x,z) \land (z \leq 15750))))$
49	$\forall x(Student(x) \rightarrow \exists y(PersonalCareCost(y) \land (requiresFundsFor(x,y) \land (y \leq 2900))$
50	$University(RhodesUniversity)$
52	$TVET(WestCoastTVETCollege)$

vocabulary resulted in the lists of classes, properties and instances of the NSFA-Sonto ontology. The classes of an ontology are the terms or concepts that describe the target domain. Some classes modelled to describe the NSFAS domain include Student, South African Student, Eligible Student, Approved Student, Denied Student, etc. In particular, the classes adopted from existing ontologies on the structure of higher education institutions and the curriculum encompass University, College, School, Department, Module, Course, etc. The finance aspect of the NSFAS domain was modelled with classes such as Study Cost, Accommodation Cost, Food Cost, Textbook Cost, Transport Cost, Tuition Fee, Fund Source, etc. A total of 49 classes were modelled for the NSFAS domain. The properties in an ontology are used to model semantic relations between the classes. A property may represent a taxonomy/inheritance relation or a predicate relation. Twenty-seven properties were used to describe the NSFAS domain. For example, the phrase "An eligible student is a South African student" that describes an aspect of the NSFAS domain, represents the taxonomy relation "is a" between the classes "Eligible Student" and "South African Student". A predicate may be a unary or binary predicate. A unary predicate represents a class assertion, that is, it represents the relation between a class and its instance, whereas, a binary predicate represents a relation between two classes of the ontology. For instance, the phrase "University of KwaZulu-Natal is a University" expresses that "University of KwaZulu-Natal" is an instance/individual of the class "University"; in this phrase, the class "University" is a unary predicate and "University of KwaZulu-Natal" its parameter; it is said that such statement is a class assertion. A number of 145 instances/individuals were collected for the NSFAS domain. The phrase "A student studies at an institution" represents a binary predicate relation "studies" between the classes "Student" and "Institution". The classes, properties and instances obtained were used to formulate 61 statements[2] in natural language that describe the NSFAS operations. The 61 statements are utilized to model the NSFASonto ontology in logic-based syntaxes next.

2.3 Logic-Based Representation of the NSFASonto Ontology

The 61 natural language statements formulated from the vocabulary of the NSFAS domain, to describe the NSFAS operations were further modelled in two logic-based syntaxes including First-order logic (FOL) and DLs, to build the axioms of the NSFASonto ontology. FOL extends the propositional logic and enables to model a knowledge domain with objects, relations/predicates and functions [24]. A part of FOL formulas representing the statements of the NSFAS domain are presented in Table 1. The left column of Table 1 indicates the indexes of the statements as in the original list of statements of the NSFAS domain. The FOL formulas in Table 1 model the following knowledge about the NSFAS domain:

– The definition of a student according to NSFAS (indexes 1 and 2),

---

[2] https://drive.google.com/file/d/1X5TKuN1F32ojt8H_Jilk3AFjLfjQruMh/view?usp=sharing.

Table 2. Part of DLs Representation of Axioms of the NSFASonto Ontology.

1	$Student \sqsubseteq Person \sqcap \exists studiesAt.Institution$
2	$SouthAfricanStudent \sqsubseteq Student \sqcap SouthAfrican$
3	$EligibleStudent \sqsubseteq (SouthAfricanStudent \sqcap \neg hasPreviousQualification \sqcap$ $(< 350000 \exists has.AnnualHouseHoldIncome))$
4	$EligibleStudent \sqsubseteq (SouthAfricanStudent \sqcap \neg hasPreviousQualification \sqcap$ $(< 600000 \exists has.AnnualHouseHoldIncome \sqcap \exists has.Disability))$
5	$EligibleStudent \sqsubseteq (SouthAfricanStudent \sqcap \neg hasPreviousQualification \sqcap$ $(< 122000 \exists has.AnnualHouseHoldIncome \sqcap < 2018 \exists startedStudying.Year))$
6	$EligibleStudent \sqsubseteq (SouthAfricanStudent \sqcap \neg hasPreviousQualification \sqcap$ $SASSAGrantRecipient)$
7	$ApprovedStudent \sqsubseteq EligibleStudent \sqcap \exists submits.ApprovedApplication$
21	$Institution \equiv (University \sqcup TVET)$
28	$ApprovedApplication \sqsubseteq \exists isApprovedBy.NSFASEmployee$
29	$DeniedApplication \sqsubseteq \exists isDeniedBy.NSFASEmployee$
39	$FundSource \equiv (Donations \sqcup ExistingFunds \sqcup GovernmentDepartment \sqcup$ $PrivateSector \sqcup LoanRepayment \sqcup SETA)$
40	$NSFAS \sqcap \exists possesses.RecoverySystem$
41	$RecoverySystem \sqsubseteq (CollectionAgency \sqcup Marketing \sqcup$ $PayrollDeuctionAgreement)$
42	$(Student \sqcap \exists studiesAt.University) \sqsubseteq$ $(\leq 7500 \exists requiresFundsFor.TransportationCost)$
43	$(Student \sqcap \exists studiesAt.University) \sqsubseteq$ $(\leq 5200 \exists requiresFundsFor.TextbookCost)$
44	$(Student \sqcap \exists studiesAt.University) \sqsubseteq$ $(\leq 15000 \exists requiresFundsFor.LivingCost)$
45	$(Student \sqcap \exists studiesAt.TVET) \sqsubseteq$ $(\leq 7000 \exists requiresFundsFor.TransportationCost)$
46	$(Student \sqcap \exists studiesAt.TVET \sqcap livingArea.Urban) \sqsubseteq$ $(\leq 24000 \exists requiresFundsFor.AccommodationCost)$
47	$(Student \sqcap \exists studiesAt.TVET \sqcap livingArea.Peri - Urban) \sqsubseteq$ $(\leq 18900 \exists requiresFundsFor.AccommodationCost)$
48	$(Student \sqcap \exists studiesAt.TVET \sqcap livingArea.Rural) \sqsubseteq$ $(\leq 15750 \exists requiresFundsFor.AccommodationCost)$
49	$(Student \sqsubseteq (\leq 2900 \exists requiresFundsFor.PersonalCareCost)$
50	$University(RhodesUniversity)$
52	$TVET(WestCoastTVETCollege)$

– The eligibility criteria for NSFAS funding (indexes 3–6); for instance, the FOL formula at the index 3 represents that an eligible student is a South African student who has not completed a previous qualification and has an annual household income of less than R350 000 (approximately 22 667 US dollars),
– The two NSFAS higher education partner institutions which are University and TVET college (index 21),
– The fact that some funding applications are approved or denied by NSFAS employees (indexes 28 and 29),
– The sources of NSFAS funding such as donations, existing funds, government departments, the private sector, loan repayments, or Sector Education and Training Authority (SETA) (index 39),
– The NSFAS recovery systems such as collection agencies, marketing, or pay-roll deduction agreements (indexes 40 and 41),
– The maximum amount of NSFAS allowance to a University student for trans-port, textbook and living costs (indexes 42–44); for instance, the FOL formula at the index 42 represents that each University student requires up to R7500 (approximately 486 US dollars) for transportation costs,
– The maximum amount of NSFAS allowance to a TVET college student for transport, textbook and living costs (indexes 45–48); as an example, the FOL formula at the index 46, represents that each TVET college student from an urban area requires up to R24000 (approximately 1554 US dollars) for accommodation costs,
– The amount of NSFAS allowance for each student personal expenses (index 49); it is modelled at the index 49 that each student requires up to R2900 (approximately 189 US dollars) for personal care costs and
– Example of University and TVET college in South Africa (indexes 50 and 52)

The FOL descriptions of the NSFASonto ontology (part in Table 1) were further represented in DLs for implementation in Protégé. The DLs representation of each FOL formula in Table 1 is provided in Table 2. The different types of statements or axioms of the NSFASonto ontology in Table 2 include various features of DLs languages such as (1) classes; they are the concepts of the NSFASonto ontology such as Student, Eligible Student, Approved Student, Institution, University, TVET, South African, etc., (2) properties also called roles; these are binary predicates that represent relationships between the classes of the NSFASonto ontology; some examples are studiesAt (index 1), submits (index 7), isApprovedBy (index 28), isDeniedBy (index 29), etc., (3) individuals which are the instances of classes in the NSFASonto ontology; for instance, Rhodes University is an individual of the class University (index 50) and West Coast TVET college is an individual of the class TVET, (4) class inclusions; they represent taxonomy relations between the classes of the NSFASonto ontology; a class inclusion is represented with the inclusion symbol; some examples of class inclusions are modelled in the axioms in indexes 1, 28 and 29, (5) restrictions; they are the constraints placed on some classes of the ontology; for instance, a South African student who has an annual household income of less than R350 000 (approximately 22 667 US dollars) is an eligible student (index 3); this statement places a restriction or constraint on the class Annual Household Income; another example states that a University student requires up to R7500 (approximately 486 US dollars) for transportation cost (index 42), which places a restriction or constraint on the class Transportation Cost. Various restrictions are modelled in the axioms at indexes 42 to 49 in Table 2.

3   Implementation

In this section, the computer and software environments used to implement the NSFASonto ontology is presented, followed by the description and discussions of the results achieved as well as the evaluation of the NSFASonto ontology.

3.1   Computer and Software Environments

The implementation of NSFASonto ontology was carried out on a DELL G5 15 laptop with the following characteristics: CPU: 2.2 GHz Intel Core i7-8750H, RAM: 16 GB, Storage: 128 GB NVMe SSD, and 1TB HDD. The DLs representation of NSFASonto ontology (part in Table 2) was further utilised to implement the ontology in OWL within Protégé version 5.5.0.

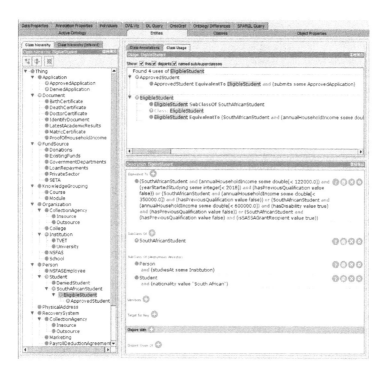

Fig. 1. Classes Hierarchy of NSFASonto Ontology in Protégé

3.2   Results and Discussions

The screenshot in Fig. 1 shows an overview of the NSFASonto ontology class structure, from the Protégé interface. The class hierarchy on the left of Fig. 1 shows the taxonomy relations in the NSFASonto ontology with 10 primary classes and 39 subclasses. In Fig. 1, the class "EligibleStudent" is selected on the bottom right and the various axioms of the ontology in which it is used are provided on the left. In particular, the top and middle right parts of Fig. 1 show the axiom of the NSFASonto ontology that defines the eligibility criteria modelled in indexes 3 to 6 in Tables 1 and 2; the eligibility criteria appear combined with the "OR" operator.

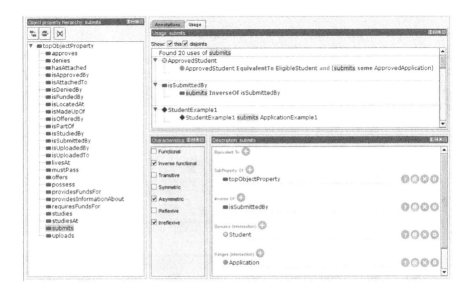

Fig. 2. List of Object Properties in the NSFASonto Ontology

Figure 2 provides a list of properties in the NSFASonto ontology. The property "submits" is selected at the bottom left of Fig. 2 and its use in the ontology is shown on the right. At the bottom right of Fig. 2, the domain and range of the property "submits" which are the "Student" and "Application" classes, respectively, are shown. The domain of a property is the class it defines, whereas, its range is its instance or occurrence. In particular, the top right of Fig. 2 shows the axiom of the NSFASonto ontology that was modelled in Tables 1 and 2 at index 7 which represents that an approved student is an eligible student who submits an approved application.

Some individuals in the NSFASonto ontology are displayed on the left of Fig. 3 including a number of Universities and TVET colleges in South Africa. The University of the KwaZulu-Natal is selected on the left and it is shown on the top right that it represents an instance of the University class. The West Coast TVET College defined in Tables 1 and 2 at index 52 can also be seen at the bottom right of Fig. 3.

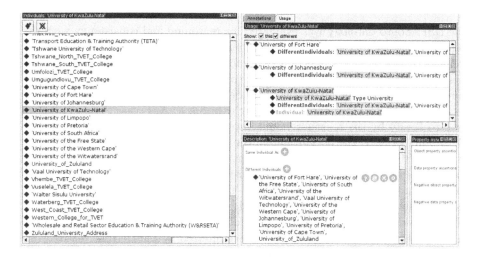

Fig. 3. Individuals of NSFASonto Ontology in Protégé

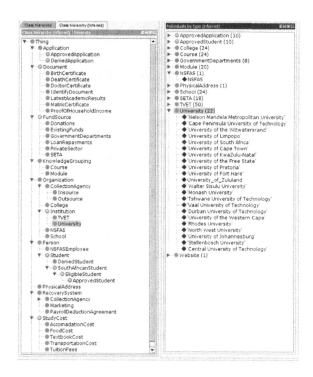

Fig. 4. Results of Consistency Check and Inference of Individuals by Type with HermiT reasoner in Protégé

3.3 Evaluation of the NSFASonto Ontology

The NSFASonto ontology was evaluated for reasoning and consistency with the HermiT reasoner in Protégé. The reasoner checks the subsumption/class hierarchy of the ontology to ascertain that each class is a subclass of another class, then, it infers a new class hierarchy of the ontology. The left side of Fig. 4 depicts the inferred class hierarchy of the NSFASonto ontology by the HermiT reasoner, which is the same as the class hierarchy in Fig. 1. Any class that would have been reclassify by the reasoner would have appeared in blue on the left side of Fig. 4. Furthermore, the background of the left side of Fig. 4 is yellow, this means that the NSFASonto ontology is consistent, that is, each class of the NSFASonto ontology can have instances. Any inconsistent class in the ontology would have appeared in red on the left side of Fig. 4. The reasoner was also used to infer all the individuals by type that are currently present in the NSFASonto ontology. The right side of Fig. 4 shows the classes of the NSFASonto ontology along with their current number of instances that were inferred by the HermiT reasoner. In particular, the University class is expanded to show its individuals, which are 22 Universities in South Africa. It is also shown other classes such as College, Course, Module, TVET, etc. with their respective number of individual that are currently recorded in the NSFASonto ontology. In a nutshell, the results in Fig. 4 show that the NSFASonto ontology is consistent and suitable for reasoning tasks.

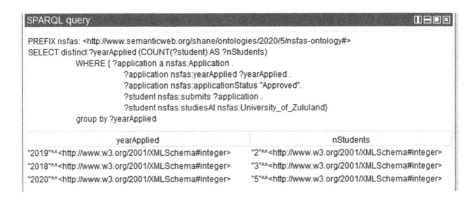

Fig. 5. Sample SPARQL Query and Outputs

The NSFASonto ontology was also evaluated with SPARQL queries. Figure 5 shows a sample SPARQL query and outputs. The query retrieves the number of approved students who were studying at the University of Zululand per year. The outputs of the query indicate that 3, 2 and 5 approved students were at the University of Zululand in 2018, 2019 and 2020, respectively. The values reported by the query are low because only few instances were loaded in the ontology for testing. This shows that the NSFASonto ontology can be used to automatically provide useful information on the state of the NSFAS financial aid in an institution. The resulting NSFASonto ontology is a formal machine readable representation of NSFAS domain that can be leveraged in web-based applications to integrate the system of NSFAS and that of public Universities and TVET colleges across South Africa, for efficient delivery of financial aid to students.

4   Conclusion

This study developed an ontology for the NSFAS operations, namely, NSFA-Sonto, in South Africa. The data was collected through secondary sources such as websites, published articles and reports as well as related ontologies, to construct the vocabulary of the ontology. The vocabulary was then utilized to design and develop the NSFASonto ontology. The ontology was evaluated with the HermiT reasoner in protégé and the results show that the NSFASonto ontology is consistent and suitable for reasoning tasks. Furthermore, the execution of SPARQL queries on the ontology yielded the expected results; this illustrated that the NSFASonto could be used to answer users' queries about the NSFAS financial aid. The resulting NSFASonto ontology is a formal machine readable representation of NSFAS domain that can be leveraged in web-based applications to integrate the system of NSFAS and that of public Universities and TVET colleges across South Africa, for efficient delivery of financial aid to students. Furthermore, the resulting NSFASonto ontology is a standard model that could be reused in other countries to build automated systems for student financial aid operations/processes. The future direction of research would be to investigate a framework that may integrate NSFAS and its partner institutions which are the public Universities and TVET colleges within South Africa, based on the NSFASonto ontology proposed in this study.

References

1. Matukane, M.M., Bronkhorst, S.: Student funding model used by the national student financial aid scheme (NSFAS) at Universities in South Africa. J. Internet Bank. Commerce (2017)
2. Writer, S.: University fees 2019: how much it costs to study in South Africa. BusinessTech. https://businesstech.co.za/news/finance/293800/university-fees-2019-how-much-it-costs-to-study-in-south-africa/. Accessed 26 May 2022
3. Revised white paper on families in South Africa. Government Gazette, 44799 (2021). https://www.gov.za/sites/default/files/gcis_document/202107/44799gon586t.pdf. Accessed 26 May 2022
4. Maphanga, C.: NSFAS has spent R4.2bn in upfront payments on students, says Nzimande. News24 (2020). https://www.news24.com/news24/southafrica/news/nsfas-has-spent-r42bn-in-upfront-payments-on-students-says-nzimande-20200326. Accessed 26 May 2022
5. Muthwa, S.: NSFAS Admin Challenges and Suggestions for Improvement (2020). https://static.pmg.org.za/190828NSFAS_FUND.pdf. Accessed 26 May 2022

6. Franke, R., Purdy, W.: Student financial aid in the united states: instruments, effects, and policy implications. In: Effects of Higher Education Reforms, pp. 153–184 (2012)
7. Ntshoe, I., de Villiers, P.: Funding sources for public higher education in South Africa: institutional responses. Perspect. Educ. **31**(4), 71–84 (2013)
8. McKay, T.J.M., Naidoo, A., Simpson, Z.: Exploring the challenges of first-year student funding: an intra-institutional case study. J. Stud. Aff. Afr. **6**(1), 19–32 (2018)
9. Busse, J., et al.: Actually, What Does "Ontology" Mean? A term coined by philosophy in the light of different scientific disciplines. J. Comput. Inf. Technol. **3**(1), 29–41 (2015)
10. Taye, M.M.: The state of the art: ontology web-based languages: XML based. J. Comput. **2**(6), 166–176 (2010)
11. Goldstein, A., Fink, L., Ravid, G.: A framework for evaluating agricultural ontologies (2019). arXiv:1906.10450. Accessed 30 May 2022
12. Domingue, J., Fensel, D., Hendler, J.A.: Introduction to the semantic web technologies. In: Domingue, J., Fensel, D., Hendler, J.A. (eds.) Handbook of Semantic Web Technologies. Springer, Heidelberg (2011). https://doi.org/10.1007/978-3-540-92913-0_1. Accessed 30 May 2022
13. Johnstone, D.B.: Higher education finance and accessibility: tuition fees and student loans in Sub-Saharan Africa. J. High. Educ. Afr. 201–226 (2004)
14. Mosikari, T.J., Marivate, H.E.: The impact of students financial aid on demand for higher education in South Africa: an econometric approach. Mediterr. J. Soc. Sci. **4**(3), 555–559 (2013)
15. Redd, K.E.: Lots of money, limited options: college choice and student financial aid. NASFAA J. Stud. Finan. Aid **34**(3), 29–39 (2004)
16. Saunders, K.: First-generation college student experience in the financial aid process. Master of Arts Dissertation. University of Nebraska, USA (2020)
17. St-Amant, P.B.: A literature review on financial student aid (2020). https://espace.enap.ca/id/eprint/272/1/2019-12-15-litt-review-student-aid-V007.pdf. Accessed 3 May 2022
18. Zemmouchi-Ghomari, L., Ghomari, A.R.: Process of building reference ontology for higher education. In: World Congress on Engineering 2013, 3–5 July 2013, pp. 1–7, London, UK (2013)
19. Ameen, A., Khan, K.U.R., Rani, B.P.: Ontological student profile. In: 2nd International Conference on Computational Science. Engineering and Information Technology, pp. 466–471. Coimbatore UNK, India (2012)
20. Chung, H., Kim, J.: An ontological approach for semantic modeling of curriculum and syllabus in higher education. Int. J. Inf. Educ. Technol. **6**(5), 365–369 (2016)
21. Noy, N.F., McGuinness, D.L.: Ontology Development 101: A Guide to Creating Your First Ontology (2001). https://protege.stanford.edu/publications/ontology_development/ontology101.pdf. Accessed 13 June 2023
22. Ye-lu, Z., Qi-yun, H., Ping, Q., Ze, L.: Construction of the ontology-based agricultural knowledge management system. J. Integr. Agric. **11**(5), 700–709 (2012)
23. Fonou-Dombeu, J.V., Naidoo, N., Ramnanan, M., Gowda, R., Lawton, S.R.: OntoCSA: a climate-smart agriculture ontology. Int. J. Agric. Environ. Inf. Syst. **12**(4), 1–20 (2021)
24. Keet, C.M.: Lecture Notes Ontology Engineering (2014). http://www.meteck.org/teaching/OntoEngLectureNotes14.pdf. Accessed 13 June 2023

The Utilization of Public-Private Partnership Frameworks in the Management of eID Projects

Stina Mander, Silvia Lips$^{(\boxtimes)}$ ⓘ, and Dirk Draheim ⓘ

Information Systems Group, Tallinn University of Technology,
Akadeemia tee 15a, 12618 Tallinn, Estonia
{stina.mander,silvia.lips,dirk.draheim}@taltech.ee
http://www.taltech.ee/en/is/

Abstract. In this paper, we examine the nature of public and private sector cooperation in the field of electronic identity (eID). We provide an overview of existing theoretical frameworks of public-private partnership (PPP) cooperation models and analyze the PPP implementation practice based on the Estonian case study. More specifically, we focus on Estonia's ID card and mobile-ID projects. During the research, we conducted five qualitative semi-structured interviews with public and private sector experts on the implementation of Estonian eID projects. This research is one of the first studies analyzing PPP theoretical framework application in the eID domain. This research identifies bottlenecks and challenges in implementing PPP projects and aims to make the process more effective. Furthermore, the study analyses the strengths and weaknesses of the PPP cooperation model implementation. Finally, we provide practical recommendations to improve the public and private sector collaboration based on the PPP framework while managing projects similar to eID.

Keywords: Public-private partnership · PPP · electronic identity · eID · project management

1 Introduction

Cooperation between the private and public sectors has received much attention from state institutions and researchers. It is necessary to find the best solutions for the needs and interests of different people, so partnership and cooperation are seen as additional resources to cope with changing societal needs [15]. However, the state must manage and finance increasing projects to ensure necessary services for citizens and residents. One solution to decrease the burden is to involve the private sector in public services provision.

This research examines the nature of public-private partnerships (PPP) and how the PPP model and its use in eID project management. eID projects were primarily chosen because countries increasingly rely on eIDs in public service

A. Kö et al. (Eds.): EGOVIS 2023, LNCS 14149, pp. 17–32, 2023.
https://doi.org/10.1007/978-3-031-39841-4_2

provision. However, due to the complexity of the eID ecosystems, the countries usually need to use private sector competence. Conducting the research is necessary because it helps the identification of bottlenecks and challenges in the implementation of eID and similar other information technology (IT)-related PPP projects and contributes to making the process more efficient.

According to World Bank's definition, "a PPP is a long-term contract between a private party and a government entity for providing a public asset or service, in which the private party bears significant risk and management responsibility"[1]. According to the National Audit Office of Estonia [17] audit, conducted concerning the cooperation between the public and private sectors in local governments there is a lack of awareness of the PPP requirements, experience in project management and implementation, and lack of sufficient analysis of whether this cooperation model is the always the most practical option. The audit emphasizes that this cooperation model needs further research to improve the PPP project management and implementation.

Infrastructure projects often use the PPP model, but it needs to be researched more in the context of IT and e-services projects. There are a lot of scientific works and studies conducted in the field of PPP, but only a few have been conducted in Estonia [13,15,18]. However, there is a clear need for more research on such cooperation opportunities in Estonia and the electronic identity (eID) area. Therefore, this research aims to find out the strengths and weaknesses of the implementation of the cooperation model of PPP to offer practical recommendations and proposals for developing this field and supporting the public and the private sector while managing PPP projects. The central research question of this study is how the PPP framework is used in the management of eID projects. We present the research results through the Estonian case study and identify how eID projects have been carried out in Estonia, what are the implications of the PPP cooperation framework, what have been the success factors of using the PPP framework in eID projects, and what are the challenges and barriers of PPP in eID projects?

The research results give us a deep understanding of the circumstances that affect cooperation in projects and achieving goals. We use triangulation of different data sources (e.g., literature, websites, and guidelines) to validate our research findings. Finally, we present the challenges and opportunities related to the PPP project and make recommendations to improve the public and private sector collaboration under the PPP.

In Sect. 2, we provide an overview of the PPP frameworks, their implementation, and related success factors. In Sect. 3, we give an overview of the research design and methodology. In Sect. 4, we introduce the Estonian case study and management of eID projects. In Sect. 5, we give an overview of the research results. In Sect. 6, we make recommendations to improve collaboration under PPP and discuss the limitations and future research perspective in Sect. 7. We conclude the research paper in Sect. 8.

---

[1] https://ppp.worldbank.org/public-private-partnership/what-ppp-defining-public-private-partnership.

2 Overview of the PPP Frameworks, Implementation and Success Factors

Several public and private sector cooperation formats are available (i.e., privatization, contractual delegation, and PPP-type of partnership). However, this particular research focuses on the application of the PPP-based model.

PPP is a contractual arrangement based on a mutual connection between the public and private sectors, and the provision of public services is jointly shared [9]. The concept of PPP has also recently been defined as the transfer of investment projects to the private sector that was previously traditionally carried out and financed by the public sector [5].

The organizational aspects of PPP can be characterized through the financial dimension (financial ties between partners) and through the organizational dimension (closeness of relations between partners) [7]. In order for PPP projects to work, clear laws and regulations on PPP projects, a strong and developed capital market, and sufficient financial strength of the private sector to invest their resources in the projects and general skill and expertise in the field of the given partnership are needed [10].

A PPP collaboration should be voluntary, and both parties should rely on each other's strengths. Resources should be optimized, and it must be beneficial for both parties. Moreover, written agreements must be in place which cover the aim, governance, and termination of cooperation [5].

The public sector is interested in PPP because they can implement high-priority projects at a quicker pace. Public institutions can rely on private companies'new technology and special management techniques, which are especially useful while carrying out complex projects. It also increases financial resources and encourages entrepreneurship [5]. The public organization withdraws itself from the regulation of its field and becomes an equal partner of the private sector organization. In a partnership, each partner has certain advantages over the other by performing certain tasks that allow the partner to bring out its best features in order to carry out projects and provide public services in the most financially efficient manner [15].

The purpose of a partnership agreement is the obligation and right of a private sector organization to upgrade, build, maintain, and/or manage an institution or system that provides a public service. The public sector retains its right based on these agreements, although the private sector invests financially in the establishment and development of a solution. The partnership also provides an opportunity to use the knowledge and skills of a private sector organization, which leads to greater efficiency. Teisman and Klijn consider that partners must jointly find the best solutions for achieving effective results, and the public sector does not dominate in this situation [22].

There are different types of PPP models, and these sub-forms differ from each other in terms of public and private sector participation, risk level, and responsibilities. For example, large infrastructure facilities are often built with the involvement of the private sector under a BOT (Build-Operate-Transfer) contract where the organizers of the construction take over the financing, organization, and responsibility of the construction and, after the construction is completed, they use, maintain and manage it for a long period. After the end of the given period, the organizers return the facilities to the state for further use [24]. The BOT model has transformed into different schemes according to the situation and need. In addition to BOT, the following abbreviations such as BOOT, DBOOT, BOOST, BRT, BLT, BTO, BOO, BBO, BT, DCMF, DBFO are also known[2]. For example, the main difference between the DBFO and BOT models is that in DBFO, the government pays an agreed monthly payment to the sponsor throughout the project, but BOT contracts are financed by the user of the facility [24]. Another common sub-form of PPP is the Design-Build-Finance-Maintain-Operate (DBFMO) contract. In this case, the private sector partner covers all phases of the project, including construction, maintenance, operation, and finding financing. The private sector makes the necessary investments and gives them to the public sector for long-term use in exchange for rent and/or service fees under agreed conditions. Therefore, it can be said that these PPP sub-forms differ from each other in terms of public and private sector participation, risk level, risk sharing, and responsibilities.

2.1 Implementation of PPP Projects

The implementation of PPP projects and cooperation with the private sector is a relatively common phenomenon in the world. Krtalic and Kelebuda have analysed that the best representatives can be considered Great Britain, USA, Japan, Australia, which are all developed countries and meet the above conditions, which are necessary for the successful implementation of a PPP project [10]. The European Union has outlined specific regulations that define the legislative framework for private sector and public sector cooperation in order to simplify its implementation. Guidelines have also been issued on how to successfully conduct PPP projects [10].

Implementation of PPP projects often varies across countries, sectors and projects. Carbonara, Costantino and Pellegrino have developed a three-layered theoretical framework in their research to analyze PPP projects [4]. The three-layer PPP analysis framework proposes three levels - country, sector and PPP project structure. Each layer is characterized by a series of dimensions and variables that are important for characterizing PPP projects.

The country layer has been divided into four dimensions which are institutional, legal, economic and financial. In this layer, Hammerschmid and Ysa have brought out that PPP task forces, legislation and government initiatives are very important to promote PPP projects and their implementation [6].

---

[2] BOOT (Build-Own-Operate-Transfer); DBOOT (Design-Build-Own-Operate-Transfer); BOOST (Build-Own-Operate-Subsidize-Transfer); BRT (Built-Rent-Transfer); BLT (Build-Lease-Transfer); BTO (Build-Transfer-Operate); BOO (Build-Own-Operate); BBO (Buy-Build-Operate); BT (Build-Transfer); DCMF (Design-Construct-Manage-Finance); DBFO (Design-Build-Finance-Operate).

Moreover, economic and financial dimensions can also describe the overall lean towards carrying out PPP method and cooperation model to deliver a certain services [4].

The sector layer consists of three dimensions. Industry organization can be characterized by regulatory regime and organizational structure which brings out the level of regulations and trends of private sector involvement/participation. Market structure helps to explain the level and elasticity of demand and givers an overview of competitors. Performance dimension is related to profitability. The project layer is divided into two big dimensions - PPP arrangement's structure and PPP arrangement's financing. This is the most detailed layer with many variables and it deals with contracts, risk allocation, resources, funding options etc. They mostly explain organizational aspects of both parties.

It is possible to say that there are many different models that the public and private cooperation can follow and they all have their specific characteristics and the degree of public sector's risk level varies but key features are still present. As PPP projects often vary across countries, sectors and projects.

2.2 Success Factors of PPP Projects

In order for PPP projects to be successful, there are certain points that should be followed. Jeffares, Sullivan, and Bovaird have indicated that the aim of the partnership must be clear and realistic [9]. Moreover, the availability of resources, both financial and human, has an important role to play in partnership. The culture of cooperation should be open, and trust must be cultivated to succeed with the project. It is often necessary to hold the motivation of employees, and they all should have a high degree of involvement to keep that kind of commitment. The roles should be clearly divided, and internal policies and defined processes help to keep the cooperation on a good level. In this case, risks and benefits must also be mapped and defined so that counter-parties know their responsibilities and what they are striving for. The functioning and performance of cooperation should be reviewed regularly and reported. This also helps to communicate with each other, set expectations, develop, and learn during the whole process [9]. There are many factors to consider when evaluating a PPP project's success. Das and Nandy suggest keeping in mind that exploring the potential, focusing on specific plans, developing concrete governance sharing structure, dealing with project planning details, supporting flexibility, and preparing a time frame are the most important aspects to achieve success [5]. Many authors bring out the importance of risk sharing in PPP projects. Risks should be allocated in a way that each partner bears the risks that they can manage the best [2]. Partnerships and projects involve uncertainties and risks, but these can be lowered by designing a thorough contract [14]. Contracts are a necessity in managing PPP projects because they create official business connections between partners [16].

Aerts has outlined an overview of the critical success factors of PPP projects [1]. They are divided into eight categories: economic, financial, legal, political, procedural, societal, structural and technical success factors. Huxman and Hubbert have brought out five success types [8]. First is achieving success and outcomes, which means making final decisions about the success of the project. The second type is getting the processes to work, which means that people do not concentrate on the final outcome, but the processes have to work as well. The third success type is reaching emergent milestones, which help to understand whether the partnership and project are going in the right way, and small achievements should be celebrated too. Gaining recognition from others and acknowledging personal pride in a partnership are other success types. These are related to personal investment into the project, and people often seek recognition to be motivated [8].

It can be said, that although the implementation of PPP projects varies across countries and sectors, evaluating the PPP project's success is very important. Researched PPP practices and success factors help the authors to make conclusions and recommendations for improving the processes and communication. Successful implementation of a PPP project is important to promote this cooperation form. Due to the complexity of PPP cooperation, there are many different risks to be shared and constantly monitored. The success factors may not be the same for all parties, so risks, benefits, and outcomes must be clearly defined and monitored throughout the project life cycle.

3   Research Methodology

This research follows the exploratory case study research methodology [23], and we use semi-structured interviews and document analysis. An exploratory study is an efficient way of understanding the situation, assessing phenomena from different angles, and presenting a new point of view [20]. Regarding the chosen research method, "how" and "why" questions are more explanatory in its nature, and they are likely to lead to the use of a case study analysis [26]. As the main research question is "How PPP framework is used in the management of eID projects?" a case study analysis was chosen. Qualitative methods, namely document analysis, and interviews were chosen as they provide wider narrative that concludes with exploratory discussion. The results of the research will be used to answer and provide an explanation of the main research questions.

Case studies analyze an existing, real-life situation and describe that situation in as much detail as possible. Two case studies will be used to analyze PPP in eID projects - procurement and project management of Mobile-ID and ID-card. As there are two projects which will be analyzed, it follows multiple holistic case study designs, namely two-case study analysis [26]. Single-case designs are more vulnerable, and having at least two cases may be beneficial from the analytical perspective. This research involves collecting qualitative data, and qualitative is "a synonym for any data collection technique (such as an interview) and data analysis procedure (such as categorizing data) that generates or uses non-numerical data" [20]. Semi-structured interviews were conducted with key actors who were involved with the project management of the two mentioned cases.

We used purposive sampling, which leads to a greater depth of information from carefully selected cases [21]. All interviewees were high-level experts or managers from the public and private sectors. Table 1 gives a detailed overview

Table 1. Interview participants.

Organisation	Time and Location
SK ID Solutions AS	27.02.23 (00:53), MS Teams
Information System Authority	28.02.23 (00:42), MS Teams
Police and Border Guard Board	01.03.23 (00:51), MS Teams
Estonian Association of Information Technology and Telecommunications	05.03.23 (00:52), On-site meeting
IDEMIA Identity & Security France SAS	16.03.2023 (00:52), MS Teams
IDEMIA Identity & Security France SAS	16.03.2023 (00:52), MS Teams

of the interview participants, interview time, and location. The main actors who were interviewed were the Information Systems Authority (ISA), Police and Border Guard (PBGB), SK ID Solutions AS (SK), IDEMIA Identity & Security France SAS (Idemia), and Estonian Association of Information Technology and Telecommunications (ITL). In this case, all relevant public sector and private sector parties, who are involved in Estonia's eID projects, are represented in the research to get an overview and analyze the full situation. A more neutral point of view on public and private partnerships was provided by the ITL expert. A semi-structured interview approach was selected with the aim of gaining insight into how previous eID projects have been carried out. Five interviews in total were carried out in February and March 2023. We interviewed six people from five different authorities/companies. Four interviews were carried out via video conferencing, and one meeting was physical. All interviews were recorded based on the participants' consent. Interviews were later transcribed manually, and it was followed by coding and analysis. Interviews were the primary source to gather data for this research. We also used secondary data sources such as academic literature, websites, and guidelines to complement the findings of the primary sources.

After the interviews were transcribed manually, the data was read carefully and analyzed in detail. The most important information was extracted from transcriptions, and key themes were identified to structure the information that was found in the research. To achieve this, we used thematic analysis [3]. The data was coded and categorized. We determined patterns to bring out case findings in a structured way. Key themes were the following words or phrases: preparation, procurement, obstacles/shortcomings, roles and tasks, communication, results, evaluation, PPP strengths, PPP weaknesses, and lessons learned.

4   Management of eID Projects: Case of Estonia

All Estonians have a mandatory and state-issued digital identity, and it is a crucial part of citizen's daily life as it provides access to different services and it allows them to sign documents. e-ID can be used via state-issued ID-card and Mobile-ID on their smartphone.

Various public and private stakeholders are involved in the Estonian electronic identity system, and cooperation in this field is significant as the service delivery depends on the cooperation of both sectors. Estonian eID ecosystem, its stakeholders, and its roles are described in various research papers [11,12,25]. Therefore, we give only a brief overview of the eID projects relevant to this study.

ID-card has been issued in Estonia since 2002, and it is the main identification document. ID-card has a chip, and it uses public key encryption to prove a person's identity in digital authentication and signing. 99% of Estonian residents have ID card[3]. A private company, SK (currently SK ID Solutions AS), started issuing ID-card e-certificates in 2002, and it has the same role today[4]. Since 2002, the Swiss company Gemalto AG (former name: Trüb AG, current name: Thales Group) has produced Estonian ID cards, but after 15 years, namely since December 2018, the French company IDEMIA Identity& Security France SAS started to deal with the production of cards. Currently, the Estonian company Hansab AS covers cards' personalization. The financing model of the ID card is quite simple. The citizen pays a state fee for getting a new ID card, and all costs are included in it. It covers the chip, poly-carbonate card, trust services, signing, and authentication. When the citizen uses the service, he or she does not pay themselves, but the e-service providers pay for it to SK.

People can use Mobile-ID with their smartphone, and it has the same functionality as an ID card, but it does not require a card reader. It is based on a SIM card, and it must be acquired from a mobile phone operator[5]. The certificates are issued by the PBGB[6]. Currently, there are three telecommunication companies in Estonia that provide SIM cards with Mobile-ID support - Telia, Elisa, and Tele2. They can all make their own requirements about the service fees and other conditions when they provide their service, but they must follow national requirements so that it works the same with different providers[7]. Mobile-ID certificates are valid for up to 5 years [19].

Regarding the historical background, Mobile-ID was created in 2007, and it received recognition as a new innovative solutions[8], but Mobile-ID, which can be used as digital identity has been issued in cooperation with PBGB since 01.02.2011. According to the SK web page, by the end of 2012, there were about 35 000 Mobile-ID users in Estonia, and by the end of 2020, there were 245 000 users. Since the beginning, there has only been one private sector provider, namely SK, who has provided Mobile-ID and cooperated in this regard with the public sector.

---

[3] https://e-estonia.com/solutions/e-identity/id-card/.

[4] https://www.skidsolutions.eu/ettevottest/ajalugu/aasta-2007.

[5] https://www.id.ee/en/mobile-id/.

[6] https://www.id.ee/en/article/mobile-id-digital-identity-document-on-smart-phone/.

[7] https://www.id.ee/en/article/applying-and-activating-mobile-id/.

[8] https://www.skidsolutions.eu/en/about/.

Users do not have to pay a state fee for the Mobile-ID[9]. Regarding service fees, the mobile operators are entitled to ask for SIM card exchange fees if the customer signs a Mobile-ID contract and the customer pays regular fees for using the service.

5   Research Results

The overall objective of this research was to explore how public and private partnership cooperates in different eID projects. This paper addresses the main research question, "how the PPP framework is used in the management of eID projects".

The results show that the interviewees have a different understanding of the PPP framework and whether it is used in Estonia or not. On the one hand, Estonia is a good example, where the country has outsourced a lot, and if we take ID card production as an example, various sections of the work have been completely handed over to the private sector. In the case of Mobile-ID, the situation is unique, it has been taken over by the public sector in the form of nationalization, and the Mobile-ID solution was not developed in cooperation between the two sectors. Therefore, it is not possible to call it a pure PPP project, although the partnership has been used after the nationalization of the solution.

The PPP framework allows sectors to focus on their strengths and competencies. The state does not maintain production or strong IT development competence in institutions, and this encourages the conclusion of partnerships. The private sector has been active in presenting its solutions and opportunities, and if the public sector creates mechanisms for structured market research and familiarization with innovative approaches, the competencies of both sectors can be merged. However, a clearer framework, internal procedures, and resources are needed for this. It is also facilitated by the organization of a competitive dialogue procurement form, but as can be seen from the cases - it requires strong legal competence and experience, and if it does not exist in the state institution, an external advisor or law firm must be procured to support the entire process with their extensive knowledge and advice.

One of the indicators of the PPP cooperation framework is long-term cooperation within one project/service. In practice, it has been the case that the Estonian state has had reliable partners, but this is rather due to the peculiarity of the small market because the competition is relatively small. For both projects or services, partners have been found within the framework of procurement, but this excludes long-term technological cooperation with a specific partner. The next new company may be the winner of the procurement, and it is still an ongoing competition.

When analyzing PPP cooperation, acting according to specific and clearly defined roles has emerged as the main concern. When it comes to cooperation

---

[9] https://www.id.ee/en/article/applying-for-mobile-id-becomes-easier-and-faster/.

between the public and private sectors, cooperation has been relatively clear in the case of Mobile-ID. Throughout history, SK has been the only partner for the state, and telecoms are subcontractors. This partnership framework between the public and private sectors is in place and long-term. In the case of the ID card project, there are more parties involved. It is more complex and interesting because the roles have also changed in the meantime, and it offers more points of analysis. Disputes have also taken place within procurement, which makes project management conditions more difficult. This leads to a situation where, in addition to new partners, the state has to consider parties with whom we have previously been in partnership, but now it is necessary to continue communication with them and settle disputes and offer the service continuously at the same time.

The research showed that the biggest bottleneck is in the public sector itself. In the form of the Mobile-ID and ID card projects are large projects related to the governance areas of different agencies and ministries, which leads to disagreements about roles. The ID-card crisis in 2017 [12] highlighted the need for clearer roles, but after the updated legislation in 2021, the updated division of roles has not been implemented in practice, according to the interviews. Therefore, we believe that the bigger conflict of PPP is not between the public and private sectors but between ISA and PBGB. This shows that cooperation between the public and private sectors, but when there are several parties in one sector, makes things difficult. The above-mentioned shows that the cooperation between the public and private sectors is formulated with contracts and clearly defined roles, tasks, and boundaries. This shows that the framework for cooperation in the private sector is relatively clearly defined (both between the public and private sectors and between the private sector and subcontractors), but if there are several parties in one sector, it can make the location of responsibility unclear. There are no contracts in the state, and the law has to be interpreted to determine roles and responsibilities, but often different parties interpret it differently.

In cooperation, it is important to act for a common goal, and in reality, all parties agree to it. If there is a common goal in mind and there is not too much focus on what is the task of one party and what is the failure of the other party, by acting together and solving problems, a better solution is reached faster. Otherwise, processes get slowed down, and instead of solving situations, problems are pushed on the shoulders of others. Open communication and the creation of different communication layers can definitely be considered a success factor. It is important to create various committees to discuss the progress of the project and solve emerging problems. This should be done at the technical level as well as at the senior management level so that the various elements are covered. An open atmosphere and the opportunity to speak and express one's views must be encouraged. A clear communication framework supports routine, regularity, and monitoring.

When analyzing PPP cooperation, the second main concern that has emerged is the contestation of procurement. Challenging a procurement requires a lot of resources from the public sector, both time and financial resources. From the point of view of procurement, it is important to take into account the contestation time, but it is extremely important to clearly outline the project's scope, requirements, and evaluation criteria from an early stage. In the case of newer and more complex procurement processes, it is worthwhile to include strong legal competence, which, for example, PGBG has also done. When it comes to roles and responsibilities, you have to rely on your strengths, and if you don't have in-house competence, you have to get it from a partner outside the organization.

In general, PPP cooperation between the parties has been assessed as good, and eID projects have been considered successful. Making changes during the existing contract on the part of the private sector is the biggest concern. This has been considered impossible on the part of the public sector, but they are expected to be more flexible. Continuous improvement of the product or service is considered important, and for this, it is also necessary to implement innovations within the ongoing cooperation. Instead, rigid contracts do not encourage this, and additional obligations are not desired.

There are certain bottlenecks in the results of the analyzed cases, and it is desired to improve their performance. For example, in the case of the ID card, it was pointed out that the chip is not being implemented to its full potential, and in that sense, it is not a fully successful project. However, the obstacle is probably not in the cooperation between the public and private sectors in the production of the ID card, which has been analyzed in this paper, but in other aspects. In the case of Mobile-ID, outdated SIM-based technology has been cited as a disadvantage, but the project as such has been successful. The upgrading of technology and the development of alternative authentication methods must come from the need, and at the moment, there are no better options mapped by the public sector.

6   Recommendations

Based on the research results and studied PPP literature, we point out recommendations that could be applied to PPP projects so that cooperation between the public and private sectors works successfully. They are based on the analysis of two eID projects, but they can also be applied more broadly to procurement in the field of public sector and information technology. Independently of the PPP model used in particular cases, every PPP project contains three main phases: preparation, implementation, and evaluation. Therefore, we divided the recommendations according to these three main phases.

Recommendations related to the project preparation phase:

– In order to encourage PPP projects, it is necessary to introduce a clear PPP framework in the public sector so that institutions follow it and are open to this form of partnership.
– Creation of PPP guidelines would help to support PPP format cooperation. For example, ITL has developed an advisory guideline for conducting procurements and the basis of the contract, which can be useful in incorporating best practices and standardizing processes.

- The public sector could consider longer-term partnerships to develop and market certain technologies. Changing partners every five years is not conducive to strong partnerships, while competition can be a driving force. It is necessary to analyze the project in question to choose the appropriate approach, but a long-term technological partnership framework could exist in the state.
- When planning the procurement, the time schedule must also take into account contestation, which can last up to a year. This requires early procurement planning. This aspect was also brought out by Das and Nandy [5].
- The state is missing the capability and mechanism to accept investments from the private sector to develop certain features that could benefit all sides. The private sector may be willing to finance some features. This mechanism could be a solution for some development issues.
- If there is a desire to develop innovative solutions, it is important to conduct market research in a structured manner before starting the procurement. This provides an opportunity to get feedback on requirements, prices, availability of providers, and other aspects. A thorough market research ensures a more successful procurement.
- Communication with market participants is also important on an ongoing basis, not only in the procurement preparation phase. This should be encouraged more in the public sector, and mechanisms put in place to take over the knowledge of the private sector. This provides an opportunity to involve the private sector at an early stage.
- Innovation and creativity must be encouraged in the public sector, but the public sector must feel a clear need to develop a solution. Just the idea of innovation is not enough. New solutions have to justify themselves, and then the corresponding steps can be taken. A feasibility analysis must be carried out.
- If there is no specific or precise knowledge of what the final solution should look like, then instead of open procurement, a competitive dialogue procurement form could be considered. Since the public sector does not have strong experience or competence in this regard, it is very important to involve strong legal support. If the requirements are clear, it is advisable to use an open procurement, in which there is more experience, but legal competence is important here as well.
- In the case of procurement, clear requirements and the scope of the procurement are important information to make an offer and analyze the commitment. Market research before announcing an open procurement helps to assess the clarity of the requirements and make corrections if necessary.
- The PPP financial model must be acceptable and beneficial to both parties. Depending on the project, this needs to be thought through so that potential partners in the private sector want to get involved.
- The public sector must carefully choose a partner so that an unfair situation does not arise. The company must be an experienced and strong partner, as eID projects are indispensable services in Estonia.

Recommendations related to the project implementation phase:

- If several parties/institutions are involved in the project from one sector, there should be one principally responsible institution so that the responsibility between different parties does not get blurred. This requires a clear negotiation of roles and the allocation of resources necessary to carry out the tasks. The importance of the clear role division was also brought out in the literature [9].
- During project management, open communication, respect between the parties, suitable management of the parties, and working towards a common goal are important. Communication must take place at different levels, and activities must be coordinated
- In order to promote innovation and successfully implement projects, the competencies of both sectors must be merged. Institutions could have clear internal procedures, frameworks, and allocated resources to support the partnership. It helps to monitor opportunities, the market, and the implementation of projects.
- Both the public sector and the private sector concentrate and rely on its competencies and strengths.
- The risks must be assessed, and the sharing of risks between the parties should be logical and equal. Defining responsibility is important. Risk allocation was also brought out as one of the PPP success factors in the literature [2].

Recommendations related to the project evaluation phase:

- After the project, the analysis of lessons learned must also be carried out in order to draw conclusions for the next procurement and cooperation project.
- Sharing experiences with other institutions and/or countries and learning from others' mistakes and success stories offers fresh insights and different perspectives.

These observations do not constitute a complete framework but provide a basis and are helpful in drawing conclusions and creating a framework.

7   Limitations and Future Research

This paper discussed the public and private sector's cooperation in eID project management. There are also some limitations with regard to this research. Firstly, procurement contracts are confidential, and cooperation is ongoing, so all interviewees were not able to fully disclose all requirements related to these projects. Also, these partnerships are still in place, and interviewees may not have been able to comment on all aspects of cooperation, as this would have impacted their work. Moreover, two case studies are still limited to making conclusive decisions and interpretations, so more IT projects could be analyzed to get a comprehensive overview of this topic.

In future work, it is important to analyze cooperation between different public sector authorities because, in the case of larger projects, there is not only one public sector authority involved from the state's side but several. Cooperation between public sector institutions was addressed in this research as well, but future research could go further and focus solely on this as this research shed light on many existing problems in this domain.

Moreover, it is also essential to set a framework for procurement and partnerships inside the public sector. Guidelines for carrying out these projects and rules of procedures in authority would be a prerequisite for successful cooperation with outside partners. Therefore, it is also important to establish a framework within the institution so that each institution can effectively fulfill its role in cooperation models. Future research could focus on creating PPP frameworks for public sector institutions.

8   Conclusion

PPP is a multi-layer collaboration format between private and public sector actors. This research explored ID-card and Mobile-ID case studies and the challenges of PPP in eID projects in Estonia. The aim was to understand how the PPP framework is used in the management of eID projects. For that purpose, we conducted five semi-structured interviews with public and private sector representatives related to the eID projects.

The research results show that the state does not maintain production or strong IT development competence in institutions, and this encourages the conclusion of partnerships. There are many different procurement forms for finding and selecting partners. Therefore, previous market research and familiarization with innovative approaches are important for the public sector to get advice from the private sector.

When analyzing PPP cooperation, acting according to specific and clearly defined roles has emerged as the main concern. When it comes to cooperation between the public and private sectors, cooperation has been relatively clear in the case of Mobile-ID. In the case of the ID card project, there are more parties involved. It is more complex and interesting because the roles have also changed in the meantime, and it offers more points of analysis. The research showed that the biggest bottleneck is in the public sector itself.

When analyzing PPP cooperation, the second main concern is the contestation of procurement. Challenging a procurement requires many resources from the public sector, both time and financial resources. It is extremely important to clearly outline the project's scope, requirements, and evaluation criteria from an early stage and include strong legal competence to carry out procurements.

Overall, eID projects in Estonia have been considered successful. The private sector's biggest concern is making changes during the existing contract. Continuous improvement of the product or service is considered important, and therefore, flexibility in this regard is advisable.

Some of the author's recommendations that could be applied to PPP projects include a clear PPP framework in the public sector, analyzing lessons learned, sharing experiences, conducting market research in a structured manner before starting the procurement, enhancing communication through different forums and layers, establishing clear requirements and scope for projects, plan enough time for accepting contestation risk. Also, PPP financial model must be acceptable and beneficial to both parties, partners must be carefully chosen, and both sectors should rely on its competencies and strengths. If several parties/institutions are involved in the project from one sector, there should be one principally responsible institution so that the responsibility between different parties does not get blurred. Open communication, respect between the parties, suitable management of the parties, and working towards a common goal are important. This research offers a view of PPP from both theoretical and practical perspectives. PPP is a good approach to solving complex issues and merging the competence, knowledge, and experience of both sectors. However, achieving good cooperation between public and private sector institutions is a complex matter that requires structured governance, collaboration, communication, and resources.

References

1. Aerts, G., Grage, T., Dooms, M., Haezendonck, E.: Public-private partnerships for the provision of port infrastructure: an explorative multi-actor perspective on critical success factors. Asian J. Shipping Logistics **30**(3), 273–298 (2014)
2. Alexandersson, G., Hultén, S., et al.: Prospects and pitfalls of public-private partnerships in railway transportation: theoretical issues and empirical experience. Int. J. Transp. Econ. **36**(1), 97–120 (2009)
3. Braun, V., Clarke, V.: Thematic analysis. American Psychological Association (2012)
4. Carbonara, N., Costantino, N., Pellegrino, R.: A three-layers theoretical framework for analyzing public private partnerships: the Italian case. Organ. Technol. Manage. Constr. Int. J. **5**(Special), 799–810 (2013)
5. Das, S.C., Nandy, M.: Public private partnership: an emerging issue. ICFAI J. Infrastruct. **6**(1), 18–23 (2008)
6. Hammerschmid, G., Ysa, T.: Empirical PPP experiences in Europe: National variations of a global concept. In: International Handbook on Public-Private Partnerships. Edward Elgar Publishing (2010)
7. Hodge, G.A., Greve, C.: The Challenge of Public-Private Partnerships: Learning from International Experience. Edward Elgar Publishing (2005)
8. Huxman, C., Hubbert, P.: Hit or myth (2009)
9. Jeffares, S., Sullivan, H., Bovaird, T.: Beyond the contract: the challenge of evaluating the performance(s) of public-private partnerships. In: Rethinking Public-Private Partnerships, pp. 166–187. Routledge (2013)
10. Krtalic, S., Kelebuda, M.: The role of the public-private partnership in providing of public goods: possibilities and constraints. In: Fifth International Conference of the School of Economics and Business in Sarajevo (ICES2010). Proceedings, pp. 1–19 (2010)
11. Lips, S., Aas, K., Pappel, I., Draheim, D.: Designing an effective long-term identity management strategy for a mature e-State. In: Kő, A., Francesconi, E., Anderst-Kotsis, G., Tjoa, A.M., Khalil, I. (eds.) EGOVIS 2019. LNCS, vol. 11709, pp. 221–234. Springer, Cham (2019). https://doi.org/10.1007/978-3-030-27523-5_16

12. Lips, S., Pappel, I., Tsap, V., Draheim, D.: Key factors in coping with large-scale security vulnerabilities in the eID field. In: Kő, A., Francesconi, E. (eds.) EGOVIS 2018. LNCS, vol. 11032, pp. 60–70. Springer, Cham (2018). https://doi.org/10.1007/978-3-319-98349-3_5

13. Lips, S., Tsap, V., Bharosa, N., Krimmer, R., Draheim, D., Tammet, T.: Management of national eID infrastructure as a state-critical asset and public-private partnership: learning from the case of Estonia. Inf. Syst. Front. (2023)

14. Lu, W., Zhang, L., Zhang, L.: Effect of contract completeness on contractors' opportunistic behavior and the moderating role of interdependence. J. Constr. Eng. Manag. **142**(6), 04016004 (2016)

15. Murumägi, M., Shak, M., Almann, A.: Public and private sector cooperation in Estonia: background and perspectives. EBS Rev. **27**, 87–103 (2010)

16. Mwesigwa, R., Bagire, V., Ntayi, J., Munene, J.: Contract completeness as a foundation to relationship building among stakeholders in public private partnership projects. Int. J. Public Adm. **43**(10), 890–899 (2020)

17. National Audit Office of Estonia: Report of the national audit office to the Riigikogu about the public-private partnerships in local authorities (2012)

18. Paide, K., Pappel, I., Vainsalu, H., Draheim, D.: On the systematic exploitation of the Estonian data exchange layer x-road for strengthening public-private partnerships. In: Proceedings of the 11th International Conference on Theory and Practice of Electronic Governance, pp. 34–41 (2018)

19. Parliament of Estonia: Identity Documents Act. Riigi Teataja (2000). https://www.riigiteataja.ee/en/eli/504072022003/consolide

20. Saunders, M., Lewis, P., Thornhill, A.: Research Methods for Business Students. Pearson education (2009)

21. Teddlie, C., Yu, F.: Mixed methods sampling: a typology with examples. J. Mixed Methods Res. **1**(1), 77–100 (2007)

22. Teisman, G.R., Klijn, E.H.: Partnership arrangements: governmental rhetoric or governance scheme? Public Adm. Rev. **62**(2), 197–205 (2002)

23. Tellis, W., et al.: Application of a case study methodology. Qual. Rep. (1997)

24. Turina, N., Car-Pusic, D.: Overview of PPP models and the analysis of the opportunities for their application. In: Proceedings of the 7th International Conference: Organization, Technology and Management in Construction, pp. 419–427. Croatian Association for Organization in Construction, Zagreb, Croatia (2006)

25. Valtna-Dvořák, A., Lips, S., Tsap, V., Ottis, R., Priisalu, J., Draheim, D.: Vulnerability of state-provided electronic identification: the case of ROCA in Estonia. In: Kö, A., Francesconi, E., Kotsis, G., Tjoa, A.M., Khalil, I. (eds.) EGOVIS 2021. LNCS, vol. 12926, pp. 73–85. Springer, Cham (2021). https://doi.org/10.1007/978-3-030-86611-2_6

26. Yin, R.K.: Case Study Research and Applications. Sage (2018)

Strategy

OntoSBehaviour: An Ontology of Students' Behaviours at Institutions of Higher Learning

Micara Ramnanan and Jean Vincent Fonou-Dombeu(✉)

School of Mathematics, Statistics and Computer Science,
University of KwaZulu-Natal, Pietermaritzburg, South Africa
fonoudombeu@gmail.com
https://www.ukzn.ac.za/

Abstract. Students who enter Universities, came from different demographics, cultural and socio-economic backgrounds. These factors contribute to the students' adjustment in the new environment and academic success. Institutions of higher education need automated systems to predict manage and monitor students' behaviours to aid academic success. However, students' behaviours in higher education institution environments have not yet been formalized to enable machine understanding for automated processing. To address this shortcoming, this study has developed an ontology of students' behaviours at institutions of higher learning, namely, OntoSBehaviour. Data was collected from secondary sources including websites, published research articles and reports, related taxonomies and ontologies, and various theories that describe human behavior. The collected data was used to formalized the OntoSBehaviour ontology in logic-based syntaxes. The OntoSBehaviour ontology was further implemented in a machine readable form in Web Ontology Language (OWL) and successfully evaluated for consistency and competency with the HermiT reasoner and SPARQL queries within Protégé. The resulting OntoSBehaviour ontology is a formal machine readable representation of students' behaviours that can be leveraged in web-based applications to provide higher education institutions with relevant data for the automatic prediction, planning, management and monitoring of students' academic activities and performance to aid their academic success.

Keywords: Student Behaviour · Ontology · Theories of Behaviour · Institutions of Higher Learning · Protégé

1 Introduction

Human behaviour is an interaction of actions, cognition, and emotions [1]. An action is all that can be seen with the eyes or measured using physiological radars. Cognitions are mental pictures and thoughts that we carry with us. Cognitions consist of various skills and knowledge, e.g. being able to use a tool correctly or sing a song. An emotion is a momentary, spontaneous conscious mental activity,

A. Kö et al. (Eds.): EGOVIS 2023, LNCS 14149, pp. 35–50, 2023.
https://doi.org/10.1007/978-3-031-39841-4_3

and a feeling that does not involve reasoning, skill, or knowledge. Emotions experienced are either positive (pleasurable) or negative (unpleasant). Actions, cognitions, and emotions do take place individually. These factors work together, which leads to an individual being able to make sense of the world and to respond. Behaviour is also an interplay between genetics (nature) and (nurture) the way we are raised in the environment [1]. Behaviour is also acquired through learning. Learning involves a process of acquiring new skills and knowledge, preferences, attitudes and evaluations, social rules, and normative considerations.

The transition from secondary to higher education can be difficult for many students who comes from a variety of backgrounds and environments and with diverse skills and knowledge. Hence, formalizing and automating students' behaviours would assist the institutions of higher learning to better plan for these transitions and improve students' chances of achieving academic success. Huber [2] investigated the reasons for student drop out and revealed the background and motivation amongst the factors that explain student drop out. Other factors identified were personal challenges such as family, money, health and friends as well as academic challenges such as not being sufficiently prepared for academics, not having sufficient academic knowledge or relevant study skills needed to address the needs of the program, full time vs part-time, wrong subject and program/university choices. Further, reasons that explain student drop out included insufficient information provided before enrolment, which may sometimes lead to insufficient focus being placed on educational and career goals, the university not being the first option, change of interest in the program or field of study, bad time management and workload demands, which lead to poor progress, dissatisfaction with the university experience and quality of curriculum content or teaching methodology. Many studies have investigated various factors that (1) impact student achievement [3], (2) contribute to student's decision to choose a University [4], and (3) influence the performance of students [5] and students' academic success [6]. All these factors influence students' behaviours in the Universities' environments as well as performance and academic success.

Institutions of higher education need to be able to predict students' reactions to various academic activities/events and students' academic performance to develop appropriate academic, screening, orientation and intervention programmes. Furthermore, managing students' behaviours on a large scale is challenging. Many students face problems which need to be addressed systematically to assist them in their academic activities. Institutions of higher education need automated systems to predict manage and monitor students' behaviours to aid academic success [5,6]. However, students' behaviours in higher education institution environments have not yet been formalized to enable machine understanding for automated processing. This shortcoming can be addressed with semantic web technologies such as ontologies, Resource Description Framework (RDF) and Web Ontology Languages (OWL). Ontology is a formal model that represents the concepts, objects and entities of a knowledge domain and the relationships between the constituents of the domain [7]. Ontology is implemented in machine readable form with standard languages such as RDF and OWL [8,9]. This enables the automatic processing and reasoning of ontology for the discovery and inference of new knowledge [10,11]. Furthermore, at this era of Big data and Machine learning, an ontology of students' behaviours could be used to store relevant data for the automatic prediction and monitoring of students' performance at institutions of higher learning. Many studies have discussed the modelling of various aspects of human behaviour with ontologies such as the affective state [12], human behaviour change [13], behaviour change intervention [14], theories of behavioural science [15] and philosophical modelling of behaviour [16]. However, none of these previous studies has focused on the development of an ontology of students' behaviours for institutions of higher learning.

To fill this gap, this study has developed an ontology of students' behaviours at institutions of higher learning, namely, OntoSBehaviour. Data was collected from secondary sources including websites [1,17–20], published research articles and reports [6,13–15], related taxonomies and ontologies [12,14–16], and various theories that describe human behaviour in general [19,20]. The collected data was used to formalized the OntoSBehaviour ontology in logic-based syntaxes. The OntoSBehaviour ontology was further implemented in a machine readable form in Web Ontology Language (OWL) and successfully evaluated for consistency and competency with the HermiT reasoner and SPARQL queries within Protégé. The resulting OntoSBehaviour is a formal machine readable representation of students' behaviours that can be leveraged in web-based applications to provide higher education institutions with relevant data for the automatic prediction, planning, management and monitoring of students' academic activities and performance to aid their academic success.

The rest of the paper is structured as follows. Section 2 conducts a review of the literature on the theories, taxonomies, terminologies and ontologies related to human behaviours. The materials and methods used to conduct the study are explained and applied in Sect. 3. Section 4 presents the experimental results and discussions. The conclusion of the paper is provided in the last section.

1.1 Related Works

Various theories and school of thoughts [17–21] explain human behaviour. According to social psychology [20], a person's thoughts, feelings, and behaviours are dictated by the social context and the individual's personality; people change their behaviour to adapt to the social context experienced. The behavioural approach founded by Pavlov [19] describes behaviour as being learnt by our experiences, association, or environmental influences. McLeod [18], an educational psychologist developed the socio-cultural theory; according to this theory, development is dependent on the interaction with people and the tools that are available to the individual by the members of the society or culture where the individual belongs. These tools, which assist the individual to make sense of the world include (1) imitative learning, where one individual tries to copy another, (2) instructed learning, which involves remembering the instructions of the teacher and then using these instructions to adjust oneself, (3) collaborative learning, which involves learning by means of a group, example is a group of peers who want to work together to learn a specific skill. The school of cognitive psychology [21] explained human cognition as the combination of all our mental capabilities such as being able to perceive, learn, remember, think, reason, and understand. According to Lu and Dosher [21], our actions are based on the mental processes of reasoning, logical thinking, memory, motivational thoughts, positive and negative thoughts. The Bronfenbrenner's bioecological model of development from the school of educational psychology, explains that an individual's development and behaviour is influenced by factors in the individual's environment [17]. The individual's environment is broken down into 5 levels including the microsystem, mesosystem, exosystem, macrosystem and chronosystem.

Bertolini et al. [3] applied the Bronfenbrenner's bioecological model to conduct a study that focused on the factors that impact student achievement and adopted a holistic approach to encourage student achievement. The authors explored student achievement factors within the five levels of the model. The microsystem level involves a person's personal traits, relationships with others such as parents and teachers, and the immediate environment. The relationship between a child and grandparents, cousins, friends and siblings are some examples at this level. Bertolini et al. [3] described the factors that encourage student learning including resilience, individual student abilities (cognitive and metacognitive), health and attendance, developmental differences (readiness for skills development), and social and moral development. At the Mesosystem level, the environment surrounding the individual or the microsystem that directly affects the student achievement was explained; this encompasses various contacts between the individuals within the microsystem such as the relationship between an individual's family and the educators or friends. Bertolini et al. [3] identified many factors that support student success including the school environment as a welcoming and safe environment for learning, parent training and partnering, professional development for teachers, leadership capacity in teachers and peer culture and achievement (expectations within the student body). The parts of the microsystem that do not influence the individual in a direct way but indirectly, constitute the exosystem; for example, if a parent were to be retrenched or experience a short time at work, this would negatively impact on the family by causing stress [17]. The Macrosystem level addresses the cultural and societal beliefs that impact a person's development such as the gender norms or religious influence [17]. Bertolini et al. [3] explained that, while factors in the mesosystem and macrosystem appear on a large scale, actions can be taken at the mesosystem level in order for the student to achieve academic success. Some of the factors in the mesosystem and macrosystem include socioeconomic disparities between families within the institutions, racism/classism, child abuse and neglect, unhealthy lifestyles. The Chronosystem comprises the transitions and shifts in one's lifespan as well as the aspect of time in the person's environment. The factors in the Chronosystem level can be either internal or external to a person's environment; an example of an external factor could be the time of a parent's death and an example of an internal factor is the physiological changes that happen when a child is growing up.

Another study [4] explored student behaviour and the factors that contribute to the student's decision on the choice of University. The study used four approaches, namely, psychological-behaviourist, social, service quality and market-oriented and service-based. The first two approaches are relevant to this study. The psychological-behaviourist perspective looks at the student's motivations and attitudes that influence student's decisions. The factors outside the individual such as the environment or influence by other individuals would not impact the student's decision. The authors [4] explained that the student's intrinsic motivation is responsible for his/her decision. In the social approach, a student is viewed as a consumer of a service. This social aspect can be seen when a teacher assigns difficult tasks and activities to the student and the student's interest in showing the ability to do better. The authors explained that the social aspects of decision making for the student are influenced by the parental goal setting and parental involvement; therefore, children's goals and career choices are influenced by that of the goals and needs of their parents [4]. The social approach also considers students' social benefits as a member of a social group such as friends, family, teachers and employees. These social groups affect the student's decisions and behaviour through the sharing of opinions and recommendations [4].

Ramesh et al. [5] used students' data and identified factors that influence the performance of students, to predict the examination results, with machine learning techniques. The study showed that parents' occupation plays a role in predicting student grades and that the type of school does not influence student performance. A similar study was carried out [6] to address the prediction of the academic success of students in institutions of higher learning. The authors reviewed the literature to identify various factors that can be used to predict students' academic success. The student related variables and the factors that impact students' performance and academic success [5,6] were adopted in the design of the OntoSBehaviour ontology proposed in this study.

The modelling of various aspects of human behaviour with ontology has been tackled [12–16]. Abaalkhail et al. [12] conducted a survey of existing ontologies for affective states and identified many factors that impact human behaviour including emotions, moods, personalities, needs and subjective well-being. The study also discussed the Abraham Maslow's Human Motivation Theory which provides a hierarchical pyramid of five needs including survival, security, safety, social, self-esteem and self-actualization. The authors indicated that the behaviour of a person depends on his level on the pyramid. A similar study [13] investigated the existing ontologies that relate to human behaviour change domain. A number of criteria where set to discriminate suitable ontologies from the list of candidate ontologies. The findings revealed that none of the fifteen ontologies discovered was related to human behaviour change, rather, these ontologies related to areas such as cognition, mental disease and emotion. A related study [14] stressed the lack of shared terminologies for describing behaviour change intervention as well as guideline and method for building ontologies of such a domain. As solution, the authors [14] proposed a method for building Behaviour Change Intervention Ontology (BCIO) as a set of ontology modules. Hastings et al. [15] proposed an ontology that represents various theories in behavioural science such as self-efficacy, control beliefs, perceived control, perceived behavioural control and empirical data annotation. The authors claimed that the resulting ontology could serve as a framework for integrating findings from the various theoretical viewpoints, for common and comprehensive knowledge as well as collaborative debate in the behavioural science domain. A philosophical discussion of an ontology of behaviour was presented by Leslie [16] who argued that behaviour should be defined functionally in terms topography and movements rather than in terms of categories.

In light of the above and to the best of our knowledge, no previous study has formally modelled students' behaviours with ontology. This is the first study that investigated existing theories, terminologies, taxonomies and ontologies related to human behaviour, to develop an ontology of students' behaviours at institution of higher learning. The resulting OntoSBehaviour ontology is a formal machine readable representation of students' behaviours that can be leveraged in web-based applications to provide higher education institutions with relevant data for the automatic prediction, planning, management and monitoring of students' academic activities and performance. The next section presents the materials and methods used to conduct the study.

2   Materials and Methods

This section presents the design of the OntoSBehaviour ontology through the building of vocabulary and logic-based modelling.

2.1   Building the Vocabulary of the OntoSBehaviour Ontology

The design of the OntoSBehaviour ontology was guided by the Ontology Development 101 (OD101) methodology [22]. OD101 provides guidelines for a rapid formalization of a knowledge domain in logic-based syntax. The OD101 has been applied to build ontologies in agricultural [23,24] and flora [25] domains. The scope of the OntoSBehaviour ontology was defined by investigating various theories, taxonomies, terminologies and ontologies related to human behaviour, through survey of published articles [2,4–6,12–16,21], research report [3], and websites' contents [1,17–20] as well as the formulation of competency questions.

As prescribed by the OD101 methodology, from all the information gathered, the vocabulary of the OntoSBeviour ontology was constituted including the classes, properties and instances. Thereafter, the vocabulary was used to formulate a number of 313 statements in natural language[1] to describe the domain of students' behaviours at institution of higher learning in detail. The natural language statements of the domain served to model the OntoSBehaviour ontology in logic-based syntax as in the next subsection.

---

[1] https://drive.google.com/drive/folders/15qATrfdm8fYqnL8QHLZeOfv3Oa5W1n0 w?usp=sharing.

Table 1. Part of FOL Representation of Axioms of the OntoSBehaviour Ontology.

	FOL Formulas
3	$\forall x(SocioCulturalBackground(x) \rightarrow \exists y(Student(y) \wedge influences(x, y)))$
4	$\forall x(PsychologicalFactor(x) \rightarrow \exists y(Student(y) \wedge influences(x, y)))$
5	$\forall x(CognitiveFactor(x) \rightarrow \exists y(Student(y) \wedge influences(x, y)))$
80	$\forall x(Emotion(x) \rightarrow PsychologicalFactor(x))$
82	$\forall x(MotivationalState(x) \rightarrow PsychologicalFactor(x))$
83	$\forall x(NegativeEmotion(x) \rightarrow Emotion(x))$
84	$\forall x(PositiveEmotion(x) \rightarrow Emotion(x))$
87	$\exists x \exists y(Interest(x) \wedge MotivationalState(y) \wedge contributesTo(x, y))$
88	$\exists x \exists y(Attention(x) \wedge MotivationalState(y) \wedge contributesTo(x, y))$
89	$\exists x \exists y(Motivation(x) \wedge MotivationalState(y) \wedge contributesTo(x, y))$
90	$\exists x \exists y(Distraction(x) \wedge MotivationalState(y) \wedge contributesTo(x, y))$
91	$\exists x \exists y(Effort(x) \wedge MotivationalState(y) \wedge contributesTo(x, y))$
92	$\exists x \exists y(Persistence(x) \wedge MotivationalState(y) \wedge contributesTo(x, y))$
93	$\exists x \exists y(Resilience(x) \wedge MotivationalState(y) \wedge contributesTo(x, y))$
94	$\exists x \exists y(Readiness(x) \wedge MotivationalState(y) \wedge contributesTo(x, y))$
103	$\forall x(Frustration(x) \rightarrow NegativeEmotion(x))$
104	$\forall x(Anxiety(x) \rightarrow NegativeEmotion(x))$
105	$\forall x(Fear(x) \rightarrow NegativeEmotion(x))$
106	$\forall x(Dread(x) \rightarrow NegativeEmotion(x))$
107	$\forall x(Diappointment(x) \rightarrow NegativeEmotion(x))$
108	$\forall x(Discouragement(x) \rightarrow NegativeEmotion(x))$
109	$\forall x(Embarrassment(x) \rightarrow NegativeEmotion(x))$
110	$\forall x(Irritability(x) \rightarrow NegativeEmotion(x))$
111	$\forall x(Loneliness(x) \rightarrow NegativeEmotion(x))$
112	$\forall x(Shock(x) \rightarrow NegativeEmotion(x))$
113	$\forall x(Boredom(x) \rightarrow NegativeEmotion(x))$
114	$\forall x(Confusion(x) \rightarrow NegativeEmotion(x))$
145	$\forall x(AddictiveBehaviour(x) \rightarrow UnhealthyBehaviour(x))$
146	$\forall x(AggressiveBehaviour(x) \rightarrow UnhealthyBehaviour(x))$
147	$\forall x(Anti-socialBehaviour(x) \rightarrow UnhealthyBehaviour(x))$
148	$\forall x(AvoidanceBehaviour(x) \rightarrow UnhealthyBehaviour(x))$
149	$\forall x(CompulsiveBehaviour(x) \rightarrow UnhealthyBehaviour(x))$
150	$\forall x(GamblingBehaviour(x) \rightarrow UnhealthyBehaviour(x))$
151	$\forall x(SuicidalBehaviour(x) \rightarrow UnhealthyBehaviour(x))$

2.2   Logic-Based Modelling of OntoSBehaviour Ontology

The 313 statements of the students' behaviour domain (see footnote 1) were further used to model the axioms of the OntoSBehaviour ontology in First Order Logic (FOL) and Description Logics (DLs). An axiom is a statement in the ontology. FOL is an extension of the propositional logic; it enables the representation of a domain logically with objects, relations/predicates and functions. Table 2 displays a part of FOL formulas for the statements of the students' behaviour domain. Each FOL formula in Table 1 is preceded by an index that corresponds to the reference number of the statement in the initial list of statements of the domain (see footnote 1). The FOL formulas in Table 1 model the following knowledge in the OntoSBehaviour ontology:

Table 2. Part of DLs Representation of Axioms of the OntoSBehaviour Ontology.

	DLs Axioms
3	$SocioCulturalBackground \sqsubseteq \exists influences.Student$
4	$PsychologicalFactor \sqsubseteq \exists influences.Student$
5	$CognitiveFactor \sqsubseteq \exists influences.Student$
80	$Emotion \sqsubseteq PsychologicalFactor$
82	$MotivationalState \sqsubseteq PsychologicalFactor$
83	$NegativeEmotion \sqsubseteq Emotion$
84	$PositiveEmotion \sqsubseteq Emotion$
87	$\exists Interest \sqcap \exists contributeTo.MotivationalState$
88	$\exists Attention \sqcap \exists contributeTo.MotivationalState$
89	$\exists Motivation \sqcap \exists contributeTo.MotivationalState$
90	$\exists Distraction \sqcap \exists contributeTo.MotivationalState$
91	$\exists Effort \sqcap \exists contributeTo.MotivationalState$
92	$\exists Persistence \sqcap \exists contributeTo.MotivationalState$
93	$\exists Resilience \sqcap \exists contributeTo.MotivationalState$
94	$\exists Readiness \sqcap \exists contributeTo.MotivationalState$
103	$NegativeEmotion(Frustration)$
104	$NegativeEmotion(Anxiety)$
105	$NegativeEmotion(Fear)$
106	$NegativeEmotion(Dread)$
107	$NegativeEmotion(Disappointment)$
108	$NegativeEmotion(Discouragement)$
109	$NegativeEmotion(Embarrassment)$
110	$NegativeEmotion(Irritability)$
111	$NegativeEmotion(Loneliness)$
112	$NegativeEmotion(Shock)$
113	$NegativeEmotion(Boredom)$
114	$NegativeEmotion(Confusion)$
145	$AddictiveBehaviour \sqsubseteq UnhealthyBehaviour$
146	$AggressiveBehaviour \sqsubseteq UnhealthyBehaviour$
147	$Anti-socialBehaviour \sqsubseteq UnhealthyBehaviour$
148	$AvoidanceBehaviour \sqsubseteq UnhealthyBehaviour$
149	$CompulsiveBehaviour \sqsubseteq UnhealthyBehaviour$
150	$GamblingBehaviour \sqsubseteq UnhealthyBehaviour$
151	$SuicidalBehaviour \sqsubseteq UnhealthyBehaviour$

- Same factors that influence student behaviour such as socio-cultural, psychological and cognitive factors (indexes 3–5),
- Some psychological factors that influence student behaviour including emotion and motivational state (indexes 80 and 82),
- Some elements that contribute to the motivational state of a student such as interest, attention, motivation, distraction, effort, persistence, resilience and readiness (indexes 87–94),
- Some negative emotions that impact student behaviour such as anxiety, fear, dread, disappointment, discouragement, embarrassment, irritability, loneliness, shock, boredom and confusion (indexes 107–114), and
- Some unhealthy behavioural traits that a student can display including addictive, aggressive, anti-social, avoidance, compulsive, gambling and suicidal behaviours (indexes 145–151),

The FOL axioms of the OntoSBeviour ontology (part in Table 1) were further represented in DLs to prepare for the implementation of the ontology in Protégé. Table 2 displays the corresponding DLs axioms for the FOL formulas in Table 1. Some features of the DLs language that are used to represent ontology are found in the axioms in Table 2 including class inclusion, class assertion, restriction and conjunction. A class inclusion is modelled in DLs with the inclusion symbol to represent binary or taxonomy/inheritance relations between ontology elements. Example of class inclusions in Table 2 are found in indexes 3–5, 80–84 and 145–151. For instance, at index 3, the class inclusion is used to represent the binary relation influences between the classes SocioCulturalBackground and Student; the axiom in index 3 can be read: socio-cultural background influences student.

Another example of class inclusion is given at index 80 to represent the taxonomy or inheritance relation between the classes Emotion and PyschologicalFactor; to mean that: an emotion is a psychological factor. A class assertion represents the instance of a class in DLs; the class is a predicate with the instance as parameter. Examples of axioms that represent class assertions in Table 2 are at indexes 103–114; specifically, the axioms at indexes 103–114 represent the occurrences/instances of the NegativeEmotion class in the OntoSBehaviour ontology. Restrictions are constraints placed on some properties of the ontology and are modelled in DLs with the existential or universal quantifier symbols. Examples of axioms with restrictions in Table 2 are at indexes 3–5, and 87–94. At index 3, a restriction is placed on the property influences to indicate that a socio-cultural background influences at least one student. The next section presents the implementation of the OntoSBehaviour ontology.

3   Implementation

This section presents the computer and software environments used to implement the OntoSBehaviour ontology, followed by the experimental results obtained and the evaluation of the OntoSBehaviour ontology.

3.1   Computer and Software Environments

The experiments were carried out on a computer with the following characteristics: Windows 10 Pro 64-bit Operating System, IntelR Core™ i5-7200U
CPU @ 2.70 GHz processor and 4.00 GB RAM. The OntoSBehaviour ontology
was implemented with the ontology editing platform Protégé version 5.5.0 and
evaluated using the HermiT reasoner.

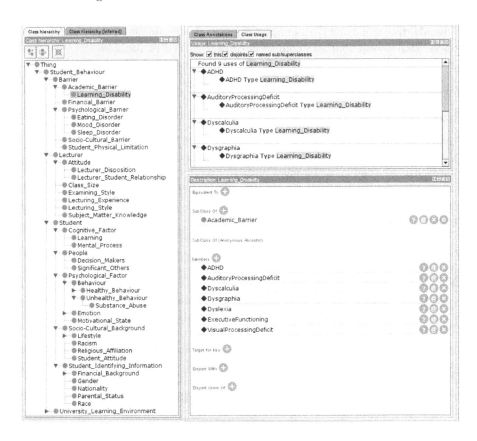

Fig. 1. Screenshot of Class Hierarchy of OntoSBehaviour Ontology in Protégé.

3.2   Results and Discussions

The screenshot in Fig. 1 depicts a partial view of OntoSBehaviour ontology
within Protégé. It is shown on the left of Fig. 1 the class hierarchy in the OntoS-
Behaviour ontology including super and sub classes. The top left part of Fig. 1
shows that the Student_Behaviour class is the top class in the OntoSBehabi-
our ontology. Student behaviour refers to the way students acts or conducts
themselves and its subclasses contain factors or contributors that influence the
student's academic success. The direct subclasses of the Student_Behaviour class

include Barrier, Lecturer, Student, and University Learning Environment. Each of these classes contain further specialized subclasses as shown in Fig. 1.

The Barrier class consists of obstacles that may prevent student's access or progress. The Lecturer class refers to an academic expert who teaches or assists students with course material and the class itself consists of lecturer related factors or contributors that influence a student. The Student class consists of factors or contributors that influence the student's behaviour and academic success. Finally, the University Learning Environment class refers to the diverse physical location, contexts, and cultures in which students learn. The class consists of factors that influence the student's academic success. In particular, the Learning_Disability class is highlighted on the left of Fig. 1 and its use and instances are provided on the right, at the top and bottom, respectively.

Fig. 2. Properties of the OntoSBehaviour ontology in Protégé.

Figure 2 displays the properties of the OntoSBehaviour ontology in Protégé. The properties are the roles or relations in the ontology. Binary relations have domains and ranges which are concepts defined by the properties and the instances or occurrences of the properties, respectively. The property influences is highlighted on the left of Fig. 2 and its domains and ranges are provided on the right. It is shown on the right of Fig. 2 the domains of the property influences which are all the things that influence student's behaviour such as the people, barrier, psychological and cognitive factors, socio-cultural background and University learning environment.

Figure 3 shows the hierarchy of subclasses of the Student class in the OntoS-Behaviour ontology generated with the OWL Viz plugin in Protégé. The subclasses of the Student class shown in Fig. 3 are various things that influence student's behaviour and the information that identify a student. The subclasses of all the class that represent the influences on student's behaviour are also provided. For instance, the subclasses of the Cognitive_Factor class are Learning and Mental_Process. The socio-cultural background elements that influence student's behaviour are also represented with the subclasses Student_Attitude, Religious_Affiliation, Racism and Lifestyle.

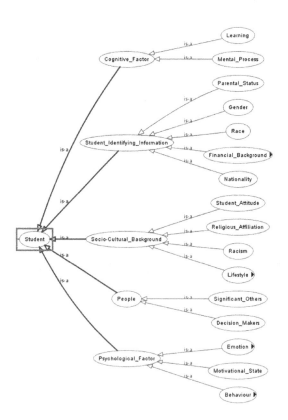

Fig. 3. Hierarchy of Subclasses of Student Class in the OntoSBehaviour Ontology.

3.3   Evaluation of the OntoSBehaviour Ontology

The HermiT reasoner was used to evaluate the OntoSBehaviour ontology in Protégé. Figure 4(a) displays the results of the use of the HemiT reasoner to infer the number of individuals by type. It is shown the Lecturer_Disposition class along with its 3 instances strict, approachable and friendly. The number of individuals per classes in the ontology are also provided in Fig. 4(a). For instance, it is shown that the Negative_Emotion has 12 individuals. The reasoner also enable to check the consistency of the ontology. Therefore, the HemiT reasoner was also invoked to perform consistency checking between the classes of the OntoSBehaviour ontology. The results of consistency check between the classes of OntoSBehaviour ontology is displayed in Fig. 4(b). The inferred class hierarchy in Fig. 4(b) is the same as the class hierarchy in Fig. 1. Furthermore, the background of the class hierarchy in Fig. 4(b) is yellow. Any inconsistencies between the classes of the ontology would have been highlighted in red in Fig. 4(b). This indicates that the OntoSBehaviour ontology is consistent.

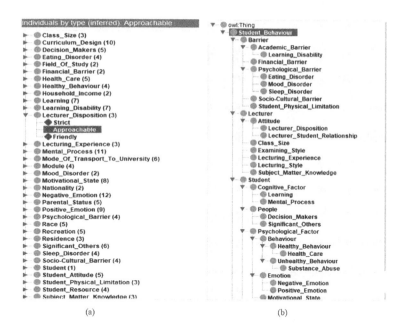

(a)   (b)

Fig. 4. (a) Inference of Individuals in OntoSBehaviour Ontology; (b) Consistency Check Between Classes of OntoSBehaviour Ontology.

Fig. 5. SPARQL Query and Outputs for Competency Question on what Behaviours Contribute to Student's Academic Success.

The OntoSBehaviour ontology was also evaluated with SPARQL queries. SPARQL queries allow users to interact with the data within the ontology as well as to check whether the ontology can answer the competency questions. Figure 5 displays the SPARQL query and outputs for competency question which asks what behaviours contribute to student's academic success. The SPARQL query in Fig. 5 retrieved from the OntoSBehaviour ontology a number of behaviours that contribute to student's academic success including art and craft activity, attention, continuous learning behaviour, effort, enjoyment, goal-oriented behaviour, etc.

4   Conclusion

This study investigated existing theories, terminologies, taxonomies and ontologies related to human behaviour in general and student's behaviour in particular, to develop an ontology of student's behaviours at institutions of higher learning, namely, OntoSBehaviour. The OntoSBehaviour ontology was successfully evaluated for reasoning tasks and consistency with the HemiT reasoner within protégé as well as for competency with SPARQL queries. The resulting OntoSBehaviour ontology is a formal machine readable representation of student's behaviours that can be leveraged in web-based applications to provide higher education institutions with relevant data for the automatic prediction, planning, management and monitoring of students' academic activities and performance. The future direction of research would be to investigate existing ontologies that model various aspects of higher education institution such the structure and organization, curriculum and syllabus, student's profile, etc. and integrate them with the OntoSBehaviour ontology, to build a global ontology of institutions of higher learning.

References

1. Fransworth, B.: Human Behavior: The Complete Pocket Guide (2022). https://imotions.com/blog/learning/research-fundamentals/human-behavior/. Accessed 03 Jan 2023
2. Huber, B.: The Role of Universities in Society. In: Liu, N.C., Cheng, Y., Wang, Q. (eds.) Matching Visibility and Performance. Global Perspectives on Higher Education. SensePublishers, Rotterdam (2016)
3. Bertolini, K., Stremmel, A., Thorngren, J.: Student Achievement Factors (2012). https://files.eric.ed.gov/fulltext/ED568687.pdf. Accessed 03 Jan 2023
4. Troisi, O., Grimaldi, M., Loia, F., Maione, G.: Big data and sentiment analysis to highlight decision behaviours: a case study for student population. Behav. Inf. Technol. **37**(4), 1–18 (2018)
5. Ramesh, V., Parkav, P., Ramar, K.: Predicting student performance: a statistical and data mining approach. Int. J. Comput. Appl. **63**(8), 35–39 (2013)
6. Alyahyan, E., Duştegor, D.: Predicting academic success in higher education: literature review and best practices. Int. J. Educ. Technol. High. Educ. **17**(3), 1–21 (2020)
7. Busse, J., et al.: Actually, what does "ontology" mean? a term coined by philosophy in the light of different scientific disciplines. J. Comput. Inf. Technol. **3**(1), 29–41 (2015)
8. Antoniou, G., Franconi, F., van Harmelen, F.: Introduction to semantic web ontology languages. In: Reasoning Web, First International Summer School 2005, pp. 1–21 (2005)
9. Domingue, J., Fensel, D., Hendler, J.A.: Introduction to the semantic web technologies. In: Domingue, J., Fensel, D., Hendler, J.A. (eds) Handbook of Semantic Web Technologies. Springer, Heidelberg (2011). https://doi.org/10.1007/978-3-540-92913-0_1. Accessed 06 Jan 2023
10. Shadbolt, N., Hall, W., Berners-Lee, T.: The semantic web revisited. IEEE Intell. Syst., 96–101 (2006). https://eprints.soton.ac.uk/262614/1/Semantic_Web_Revisted.pdf. Accessed 06 Jan 2023
11. Taye, M.M.: The state of the art: ontology web-based languages: XML based. J. Comput. **2**(6), 166–176 (2010)
12. Abaalkhail, R., Guthier, B., Alharthi, R., El Saddik, A.: Survey on ontologies for affective states and their influences. Semant. Web **9**(4), 441–458 (2018)
13. Norris, E., Finnerty, A.N., Hastings, S.G., Michie, S.: A scoping review of ontologies related to human behaviour change. Nat. Hum. Behav. **3**(2), 164–172 (2019)
14. Wright, A.J., et al.: Ontologies relevant to behaviour change interventions: a method for their development. Wellcome Open Res. **5**(126), 1–32 (2020)
15. Hastings, J., Michie, S., Johnston, M.: Theory and ontology in behavioural science. Nat. Hum. Behav. **4**, 226 (2020)
16. Leslie, J.C.: The ontology of behaviour. Eur. J. Behav. Anal. **20**(2), 160–165 (2019)
17. Bronfenbrenner: Bronfenbrenner's Bioecological Model of Development (2017). https://learning-theories.com/bronfenbrenners-bioecological-model-bronfenbrenner.html. Accessed 07 Jan 2023
18. McLeod, S.: Lev Vygotsky's Sociocultural Theory (2018). https://www.simplypsychology.org/vygotsky.html. Accessed 07 Jan 2023
19. Walinga, J.: Behaviourist Psychology (2014). https://opentextbc.ca/introductiontopsychology/chapter/2-3-behaviourist-psychology/. Accessed 07 Jan 2023

20. Spielman, R.S., Jenkins, W.J., Lovett, M.D., Dumper, K., Lacombe, A.: Psychology. OpenStax (2020). https://opentextbc.ca/psychologyopenstax/. Accessed 07 Jan 2023
21. Lu, Z., Dosher, B.A.: Cognitive psychology. Scholarpedia **2**(8), 2769 (2007)
22. Noy, N.F., McGuinness, D.L.: Ontology Development 101: A Guide to Creating Your First Ontology (2001). https://protege.stanford.edu/publications/ontology_development/ontology101.pdf. Accessed 13 Jan 2023
23. Ye-lu, Z., Qi-yun, H., Ping, Q., Ze, L.: Construction of the ontology-based agricultural knowledge management system. J. Integrat. Agric. **11**(5), 700–709 (2012)
24. Fonou-Dombeu, J.V., Naidoo, N., Ramnanan, M., Gowda, R., Lawton, S.R.: OntoCSA: a climate-smart agriculture ontology. Int. J. Agric. Environ. Inf. Syst. **12**(4), 1–20 (2021)
25. Panawong, J., Kaewboonma, N., Chansanam, W.: Building an ontology of flora of Thailand for developing semantic electronic dictionary. In: 3rd International Conference on Applied Science Technology (ICAST 2018), pp. 1–7 (2016)

Public Health Data and International Privacy Rules and Practices: A Case Study of Singapore

Ali Alibeigi[1,2](✉) [iD] and Abu Bakar Munir[1] [iD]

[1] Faculty of Law, University of Malaya, Kuala Lumpur, Malaysia
`{alii,abmunir}@um.edu.my`
[2] Faculty of Law and Humanities, Isfahan (Khorsagan) Branch, Islamic Azad University, Isfahan, Iran

Abstract. States pursued many operative measures to fight Covid-19 disease like "Contact Tracing" Apps. With the recent amendments to the law, the Singapore police could obtain any data, including information gathered by the contact tracing app and wearable token to facilitate criminal probes. This study aims to analyze rules and standards concerning personal data processing and the privacy policy of the TraceTogether App from both domestic and international perspectives. Doctrinal legal study using explanatory and comparative approaches was applied. This study examined and evaluated important international and regional regulations and guidelines concerning the privacy of contact tracing apps in fighting pandemic. The police can use personal data collected by the 'TraceTogether' App to investigate serious crimes. The results reveal that police access to contact tracing data for irrelevant purposes is a broken promise and eradicates individuals' trust in the future. The findings show that Singapore police access to personal data is against international standards and regulations on data protection, although the Singapore Personal Data Protection Act does not apply to public sectors like Police. The results highlighted that while public health is important, privacy, ethics and human rights must be observed at the same time. Singapore's government is the "custodian" of the contact tracing data, and "stringent measures" should be established to safeguard the personal data. The results are noteworthy for lawmakers since the urgent codification done by the Singapore government to legalize such access is against principles of the personal data protection law, ethics, international rules, norms, and practices.

Keywords: TraceTogether · Compliance · Covid-19 · PDPA · Clinical Ethics · Data Security · Health Data · Health Regulations · Personal Data Protection · WTO

1 Introduction

By the end of 2019, a novel coronavirus disease called "Covid-19" spread out in China and soon after, the entire world. International Health Regulations (2005) has obliged member states to develop public health surveillance systems to collect data and fight

A. Kö et al. (Eds.): EGOVIS 2023, LNCS 14149, pp. 51–66, 2023.
https://doi.org/10.1007/978-3-031-39841-4_4

pandemics. Hence, states pursued many operative measures to fight this highly contagious disease. One of the efforts is "Contact Tracing" mostly through mobile Apps to inform persons who have been in proximity to a person who has tested positive for Covid-19. The App works through Bluetooth or GPS when for example A and B smartphones are nearby [15]. Countries adopted different apps and methods like Bluetooth, GPS, digital online questionnaires, and Mobile networks [4]. However, privacy and confidentiality are regarded as salient concerns of these apps while many questions remain like the types and amount of collected data, retention period, who can access the data, individual access rights, etc. [15].

Singapore was one of the first few nations to launch its contact tracing app; "TraceTogether". In January 2021, it was announced that TraceTogether personal data will be accessible to police for criminal investigations. After the criticism increased, it was explained that access is restricted to serious crimes. Subsequently, a Bill was passed on 2 February 2021 urgently to legalize such access. The question arises whether personal data collected for a specific purpose of an emergency can be shared with third parties for an irrelevant purpose. This study aims to examine the data sharing by the Singapore "TraceTogether" app from the international information privacy regulation and guidelines perspectives. The study applied doctrinal legal study through explanatory and comparative legal approaches. International and regional documents and standards, organizations' Guidelines, and domestic laws have been examined and analyzed.

2   Singapore Contact Tracing System

Singapore launched a Quick Response (QR) Code as a part of the SafeEntry contact tracing system. Individuals who enter a public place scan the code and register by their name, ID number, and contact number. On 20 March 2020, "TraceTogether" App (hereinafter App) was released. Later, a small digital fob called "TraceTogether Token", was rolled out on 28 June 2020. When two users like A and B are nearby, their encrypted IDs exchange through Bluetooth. The App helped the authorities to track down individuals who were close to an infected person in different places.

2.1   TraceTogether Privacy Policy

TraceTogether Privacy Safeguards (Privacy Statement), includes both TraceTogether App and the TraceTogether Token. According to the statement, they collect 'contact/mobile number, identification details, and a random anonymized user ID'. It elucidates that "The Bluetooth data exchange does not contain any personal identifiable information". It highlights that the exchanged Bluetooth data between devices are anonymized temporary IDs and will be deleted automatically after 25 days. The individuals' right to deletion of their data from the server and the method was recognized. It specifies once the contact tracing ceases, the collected data will be deleted. The purpose of the collection is to inform any individual who was in proximity to an infected person. The statement highlights "Data that is shared with the Ministry of Health (MOH) will be used for COVID-19 contact tracing". However, as a sub-purpose, for the safety and security of citizens, authorized police officers may invoke the Criminal Procedure Code

(CPC) powers and ask the users to upload their App data for criminal investigations. The sub-purpose was included on 4 January 2021. It was cited that the changes include clarification on "how the Criminal Procedure Code applies to all data under Singapore's jurisdiction". Later, on 1 February 2021, more modifications were made to this section. The following sentence is omitted: "powers to request users to upload their TraceTogether data" and a general statement is replaced "Any data shared with MOH can only be used for the purpose of contact tracing, except when there is a need to use the data for criminal investigation and proceedings". There was an ambiguity with the phrase "criminal investigations". Later, it was defined "criminal investigation and proceedings relating to seven categories of serious offences" has been added. A question arises as the access is only to the individuals' App data (available at App) or access to the whole MOH data center. Furthermore, if the systems for all contact tracing efforts will be closed at the end, then why the privacy statement was amended and subsequently a Bill was passed urgently? While the privacy statement was easily accessible on the website and written in simple language, it seems to be vague and incomplete. The contact information for inquiries (except deletion) was not provided. The statement kept silent on individuals' right to access or update their data.

2.2 PDPA'S Inapplicability

Contrary to the international approach, the Singapore Personal Data Protection Act (PDPA) does not apply to the public sector. The rules concerning the processing of personal data by public entities in Singapore are governed by the Government Instruction Manual on Infocomm Technology & Smart Systems Management (IM on ICT&SS Management) and Public Sector (Governance) Act 2018 (PSGA) which entered into force on 1 April 2018 along with other specific legislations. The Singapore Personal Data Protection Commission Office (PDPC) issued "Advisories on Collection of Personal Data for COVID-19 Contact Tracing and Use of SafeEntry" providing guidelines for organizations collecting data through SafeEntry [21]. It provided that collection, use, and disclosure of relevant personal data by organizations is allowed during the fight against Covid-19 for contact tracing and other response measures.

2.3 Legal Basis for Police Access to Data

Smart Nation and Digital Government Office (SNDGO) in January 2021, stated that the government decided to pass a law concerning the use of contact tracing data for the criminal investigation of serious offenses in favor of the public interest [23]. The Government decided to pass the law urgently in February 2021 [6]. It will legalize the use of contact tracing data for seven categories of crimes like terrorism, murder, kidnapping, and the most serious drug trafficking cases. The issue was explained by the Minister for Home Affairs Desmond Tan in a Parliamentary session on 4 January 2021 and on the same day, the Privacy Statement on the TraceTogether website was updated adding the use of data for criminal investigations [18].

The police power to access App data is justified based on Section 20 of the Criminal Procedure Code (CPC). According to Singapore Foreign Minister Dr. Balakrishnan police had accessed TraceTogether data before, in a murder investigation [24]. Allowing the police to access contact tracing data was "legalized" under the CPC and the Covid-19 (Temporary Measures) (Amendment) Act 2021. However, individuals had essentially consented to use surveillance technologies based on the privacy statement that has restricted the purpose of data collection merely to the Covid-19 pandemic. The government amended the privacy statement of the App after nine months in operation and a huge amount of data had been collected. Furthermore, it is against the principles of data protection law as well as international rules, norms, and practices.

3   International Norms and Practices

There are critical legal and ethical dimensions to the use of Digital Contact Tracing and Quarantine (DCTQ) tools, including privacy and surveillance, which differ depending on the technology being used and its application. These tools should be governed by effective legal frameworks grounded in sound ethical principles to ensure their use is lawful, proportionate, and properly managed. Doing so will foster public trust, acceptance, and voluntary uptake, enhancing the effectiveness of selected tools as well as overall public health efforts [28].

3.1   United Nations

According to the UN High Commissioner for Human Rights Michelle Bachelet, openness and transparency are key factors to encourage people's participation in health-protecting measures. She tried to convince the governments that human rights concerns must be the core basis for any response to Covid-19 [3]. The UN Privacy Policy Group issued a joint statement on *Data Protection and Privacy in the COVID-19 Response* on 19 November 2020 [25]. Being in line with the UN Personal Data Protection and Privacy Principles, it was endorsed by the UN System Organizations. The statement refers directly to "digital contact tracing" and the necessity of personal data to combat the pandemic. It states specifically the risk of collecting a huge amount of data by digital contact tracing and use for purposes not related to the Covid-19. The statements clarify that human rights, privacy, and data protection principles must be respected (Table 1). Moreover, lawful and legitimate purposes, limited scope and retention period, security, deletion, and transparency must be properly considered.

The United Nations Secretary General in the call to action for *COVID-19 and Human Rights: We are all in this together* offers 6 key messages to be considered and centered in any response to the pandemic [27]. The fifth is "threat is the virus, not the people" which clarifies that "emergency and security measures if needed, must be temporary, proportional and aimed at protecting people". The report refers to respecting the rule of law in responses like states of emergency, emergency legislations, excessive use of force, and of use of surveillance and tracking technology to collect personal data "in ways that are open to abuse to prevent politically driven measures" (Table 1). It recommends providing safeguards while using new technologies including "purpose limitations and adequate privacy and data protections". It elucidates that new technologies have a high potential for abuse. Hence, the responses must protect individuals' data through legitimate public health objectives, time-bound, lawful, and proportionate [27].

Along the same lines, the UN Special Rapporteur on the promotion and protection of the right to freedom of opinion and expression David Kaye formulated six principles that should govern surveillance during the pandemic (Table 1):

1. Any authorization of surveillance should be contained in precise and publicly accessible laws and only be applied when necessary and proportionate to achieve a legitimate objective (such as protecting public health);
2. Authorization of surveillance of specified individuals should be based on independent evaluation, preferably by a judicial authority, with appropriate limitations on time, location, manner, and scope;
3. Rigorous record-keeping should be required so that individuals and oversight bodies can ascertain that surveillance was conducted for legitimate public health purposes;
4. Principle 4 explicitly emphasizes that any personal data collected should be subject to strict privacy protections to ensure against disclosure of personal information to anyone not authorized for public health purposes;
5. Some personal data should be expressly excluded from collection, such as the content of a person's communications, and robust safeguards must be put in place to ensure against any government or third-party misuse of such data, including use for purposes unrelated to public health emergency; and
6. Where personal data is anonymized, the State and any third-party actor involved in the collection must be able to demonstrate such anonymity.

The above-discussed documents specify that human rights principles and personal data protection must be the first concern in fighting the epidemic. They refer to the risks of the applied measures like the pursuit of political intentions, security considerations, excessive data, irrelevant purposes, emergency legislations, excessive use of force, and even the use of surveillance technologies itself. Therefore, they recommend certain principles to be observed by both public agencies and private such as transparency, openness, the proportionality of the measures, legitimate and limited purposes, retention period, security safeguards, and deletion of the data center to protect the privacy and personal data of individuals.

3.2 World Health Organization

World Health Organisation (WHO) and its member states have adopted the International Health Regulations (IHR) to prevent, protect against, control and provide a public health response to the international spread of disease (Table 2). The IHR (2005) is a legally binding instrument. Article 13 requires state parties to develop, strengthen and maintain the capacity to respond promptly and effectively to public health risks and public health emergencies of international concern. Surveillance is obligatory on the state party under Article 5 which clarifies the state parties shall develop, strengthen and maintain the capacity to detect, assess, notify, and report events as soon as possible but no later than five years from the entry into force of the Regulations. Article 45 provides that health information collected or received by a state party from another state party or from WHO which refers to an identified or identifiable person shall be kept confidential

Table 1. United Nations Standards on Contact Tracing Data.

No	Document	Requirements
1	Data Protection and Privacy in the COVID-19 Response, November 2020	To respect human rights, privacy, and data protection principles, lawful and legitimate purposes, limited scope, retention period, security, deletion, transparency
2	UN High Commissioner for Human Rights, Michelle Bachelet	Human dignity and rights, openness and transparency
3	UN Secretary General call to action for COVID-19 and Human Rights: We are all in this together, April 2020	Priority to protection of people's lives, no discrimination but an equitable and universal response, all people must involve in response, not threatening people by temporary and proportional security measures, a collective universal response
4	UN Special Rapporteur on the promotion and protection of the right to freedom of opinion and expression, David Kaye	Surveillance under laws if necessary for a lawful objective, surveillance of specified individuals considering limitation on scope, time, method & location, meticulous record-keeping to prove legitimate surveillance for public health purposes, forbidden of data sharing to unauthorized persons by observing personal data protection laws, no excessive data collection, no irrelevant purposes, authority ability to demonstrate anonymized data

and processed anonymously as required by national law. However, state parties may disclose and process personal data when essential for assessing and managing public health risks. This has to be done under national law, and WHO must ensure that the personal data are: (a) processed fairly and lawfully, and not further processed in a way incompatible with that purpose; (b) adequate, relevant, and not excessive; (c) accurate and, where necessary, kept up to date; every reasonable step must be taken to ensure that data which are inaccurate or incomplete are erased or rectified, and (d) not kept longer than necessary.

The WHO has addressed the necessity of the right to privacy for the whole process of the *contact tracing system under Contact tracing in the context of COVID-19 interim guidance* issued on 10 May 2020 [30]. It provided standards and definitions of contact tracing tools. Further, it offers guidelines on data protection (Table 2). The interim guidance issued in May 2020 was later updated on 1 February 2021. *Ethical Considerations to Guide the Use of Digital proximity tracking technologies for COVID-19 contact tracing Interim Guidance*, 28 May 2020 directly address the governments to protect individuals' right to privacy and ethical concerns [31]. It states that the "use of personal data may threaten the fundamental human rights and liberties during and after the Covid-19 pandemic". It added that now the role of law, policies, and limitations on the use of digital proximity tracking technologies are significant. It provided seventeen principles to be observed [31]. The time limitation principle recommends that the use of measures are temporary and limited and after the end of Covid-19 all measures must be fully shut down. The data minimization principle requires a limited amount of data collection, limited processing, and retention. Under the use restriction principle, the data should not be used for commercial and advertisement.

Furthermore, it specifies that while the governments may have their own data protection laws and frameworks, sharing of data with government bodies, third parties, law enforcement, or immigration departments that are not engaged with public health response must be prohibited [31]. The voluntariness principle states that the use of applications should be voluntary and informed and individuals are free to shut down the application. The transparency and explainability principles specify that data collection and processing should be transparent by providing a clear purpose of collection, types of data, method of storing, sharing, and retention period. The limited retention principle requires the deletion of the server at the end of the pandemic. Another document, *WHO guidelines on ethical issues in public health surveillance* emphasizes a clear objective and ethics (ethical surveillance) to be the core concern of the process [29]. It recommends that the surveillance data be collected only for "legitimate public health purpose". Transparency, risk management (prior and after), and security of data must be accomplished. The guidelines has pointed to arrest, prosecution, death penalty, and expulsion as the potential harms of disclosure of public health surveillance data [28]. Guideline 17 specifies the prohibition of sharing identifiable surveillance data with agencies that are likely to use data to take action against people or process data for unrelated purposes. Sharing of data is obligatory with other national and international health agencies through protective measures and justification. Nonetheless, the collected data can be shared and used by public health agencies only for research purposes.

Considerations for strengthening legal frameworks for digital contact tracing and quarantine tools for COVID-19 – Interim Guidance, was published on 15 June 2021 by the WHO Western Pacific Region [28]. It aims to support the Member States in the Western Pacific Region to review, develop and monitor their legal frameworks for DCTQ tools, guided by the ethical principles for digital proximity tracking technologies suggested by the WTO.

WHO states that DCTQ tools generally involve the collection and use of information from or about individuals. This presents an incursion into private and social life, impacting the right to privacy and raising issues of data protection and control. The information DCTQ tools utilize is often highly sensitive and capable of being misused, including information about the movements and health status of individuals and communities. Seemingly benign information including so-called metadata may also identify individuals and reveal personal information that is sensitive in a given context [28].

The organization recommends that clear legal frameworks, preferably in the form of legislation, are needed to address the legal and ethical dimensions of DCTQ tools and ensure their use is lawful, proportionate, and in the interest of public health (Table 2). Legal frameworks must be in place to govern the use of personal information and other data to ensure that privacy is respected and to prevent loss, unnecessary intrusion, and commercial exploitation.

This will help to secure public trust and acceptance of DCTQ tools, encouraging voluntary uptake and usage, which are critical to the effectiveness of many tools. Failing to assure the public that data are adequately protected and not able to be used for purposes other than public health could also have broader implications for trust and participation in public health efforts [28]. WHO is concerned about the risk of DCTQ tools being used for purposes outside of public health and beyond the COVID-19 pandemic, including broader population surveillance and law enforcement. Furthermore, the involvement of private actors in the development and implementation of these tools also increases the potential for data to be extracted and processed for commercial purposes, as well as for public health infrastructure to build a dependency on commercial products.

WHO provides a non-exhaustive list of legal and ethical considerations for member states to assess their legal frameworks in developing the DCTQ tool. For example, under the use restrictions, three questions need to be asked: (i) Does the legal framework limit access to data collected by the tool? (ii) Does the legal framework limit the use of data for public health and related purposes? (iii) Does the legal framework restrict the use of the data for other purposes, including law enforcement and commercial activities? WHO as an international specialized health agency has developed different recommendations on technical and medical issues to run the contact tracing system. However, it restates that privacy and data protection concerns must be respected during the surveillance process. WHO is concerned about the threat to human rights and liberties, sharing of data with unauthorized government agencies or law enforcement bodies which may lead to arrest, prosecution, the death penalty, and expulsion of individuals. It elucidates that it is necessary to observe proportionality and transparency, accuracy and accountability, risk management, specified and limited data collection, limited and specified purposes, retention period and time limitation, prohibition of sharing data with unauthorized government bodies or third parties, explainability, security safeguards, and independent surveillance.

3.3 European Union

The European Commission has published a document named *Guidance on Apps supporting the fight against COVID 19 pandemic in relation to data protection* [13] (Table 3). To ensure a coherent approach across the EU, the document sets out features and requirements which apps should meet to ensure compliance with the EU privacy and personal data protection legislation in particular the General Data Protection Regulation (GDPR) and the ePrivacy Directive. The document outlines certain elements for a trustful and accountable use of apps and aims to guide how to limit the intrusiveness of the app functionalities.

The Guidance states that the legal basis (Union or Member State law) should provide for the processing. The purpose should be specific so that there is no doubt about what kind of personal data is necessary to process to achieve the desired objective explicitly. There may be several purposes for each functionality of an app. To provide individuals with full control of their data, the Commission recommends not bundling different functionalities. In any event, the individual should have the possibility to choose between different functionalities pursuing each with a separate purpose [13]. The document explicitly states, "The Commission advises against the use of the data gathered

Table 2. World Health Organisation Standards on Contact Tracing Data.

No	Document	Requirements
1	International Health Regulations, 2005	To build a minimum set of core capacities to protect citizens; obligatory surveillance; develop and maintain the capacity to respond effectively to public health risks and emergencies of international concern; promptly detect, assess, notify and report the events no later than five years from the entry into force of the IHR, health data collected by a state party from another state, referring to an identified or identifiable person shall be kept confidential and processed anonymously under national law. WHO must ensure: fairly and lawfully processing of data and in line with purpose; adequate, relevant and limited data; accurate and up to date data; retention period (d) not kept longer than necessary
2	Contact tracing in the context of COVID-19 interim guidance, May 2020	Guarantee the privacy and personal data protection under the domestic legal frameworks, to observe the ethical principles and privacy in handling the data, processing, storage and handling must communicated to those concerned in a clear way, to assess the contact tracing tools in advance to ensure the data protection standards
3	Ethical considerations to guide the use of digital proximity tracking technologies for COVID-19 contact tracing interim guidance, May 2020	Time limitation, Testing, and evaluation, Proportionality, Data minimization, Use restriction, Voluntariness, Transparency, and explainability, Privacy-preserving data storage, Security, Limited retention, Infection Reporting, Notification, Tracking of COVID-19-positive cases, Accuracy, Accountability, Independent oversight, Civil society, and public engagement
4	WHO guidelines on ethical issues in public health surveillance	Clear objective for processing; ethical surveillance for "legitimate public health purpose"; transparency; risk management (prior and after), security

(continued)

Table 2. (*continued*)

No	Document	Requirements
5	Considerations for strengthening legal frameworks for digital contact tracing and quarantine tools for COVID-19 – Interim Guidance, June 2021	Develop clear legal frameworks preferably a legislation to address legal & ethical dimensions of DCTQ tools; DCTQ use lawfully, proportionate in interest of public health; prevent loss, unnecessary intrusion, and commercial exploitation; proportionality; time limits; data minimization; transparency & explainability; voluntariness (consent); use restrictions; security; notification; governance and oversight; human rights and equity

under the above conditions for other purposes than the fight against COVID-19. Should purposes like scientific research and statistics be necessary, they should be included in the original list of purposes and clearly communicated to users".

The European Data Protection Board (EDPB) in response to the Head of Unit European Commission DG for Justice and Consumers, stated that they have provided guidelines and will support the issue with future guidelines [12]. The EDPB emphasizes that the obligation for data protection and respect of human rights is not only a legal obligation but a "requirement to reinforce the effectiveness of any data-based initiatives" to fight the Covid-19 epidemic. It states that the use of contact-tracing technologies should be voluntary. It recognizes the necessity of data protection impact assessments, privacy by design, and privacy by default mechanisms [12]. The EDPB draws attention that the system must be shut down at the end of the crisis and the stored data must be deleted or anonymized. According to Andrea Jelinek, the Chair of EDPB, the General Data Protection Regulation (GDPR) has provided rules for the processing of data to combat Covid-19 [12]. Under the GDPR, employers and public health authorities can process personal data without prior consent since it is necessary for "reasons of public interest in the area of public health, to protect vital interests or to comply with another legal obligation". The *Statement on the processing of personal data in the context of the COVID-19 outbreak* published by the EDPB clarifies that under emergency conditions, limitations of individuals' freedom are legitimate if it is proportionate and only for the emergency period [9]. It suggests that the processing of data under emergency conditions must be lawful, in a sense under union law and national laws. It explains that the processing of special categories of personal data by competent public authorities and employers is allowed according to Articles 6, 9, and Recital 46 of the GDPR to respond to a pandemic. It draws the attention that clear, specific, and explicit purposes of the processing and retention period must be notified to the individuals in clear and plain language through easily accessible means. The statement highlights that the collected data should not be shared with unauthorized parties (Table 3).

The EU toolbox for mobile applications for contact tracing and warning is a mechanism developed by the EU members in collaboration with the e-Health Network, supported by the European Commission (EC) concerning contact tracing mobile applications in the EU to combat Covid-19 [10]. In line with this mechanism, the EC supported the use of digital technologies and data to fight against Covid-19 but defends the people's freedom, single market, fundamental rights, and particularly the right to privacy and protection of personal data [11]. The Toolbox affirms that GDPR governs the processing of personal data including health data by obtaining explicit consent, and if the processing is necessary for monitoring, alert, and prevention of communicable diseases and health threats. The Toolbox refers to the European Directive 2011/24/EU [8], which requires EU eHealth Systems to achieve a great level of trust and security, safe and quality healthcare. Furthermore, the Toolbox restates that functions of mobile applications may affect the "exercise of certain fundamental rights such as *inter alia* the right to respect for private and family life". Hence, any limitation to the exercise of any fundamental right through the laws of the state must be under general principles of Union law (according to Article 6 of the Treaty on the European Union), domestic constitutional traditions, and also rules of international law.

According to Toolbox, the processing of data by both public health authorities and research institutions must be in line with the principle of data minimization, in the sense the processing of personal data is allowed only where adequate, relevant, and limited to what is necessary, providing adequate safeguards like "pseudonymisation, aggregation, encryption, and decentralization" [8] (Table 3). Furthermore, cybersecurity and data security are crucial to secure the availability, authenticity, integrity, and confidentiality of collected personal data. The document makes an instructive and practical suggestion that is consultation with data protection authorities who are fully aware of the EU laws and regulations on the protection of personal data. Recommendations while addressing the process for developing a Toolbox restate the protection of fundamental rights, privacy, and data protection, purpose limitation only to fight with a pandemic, and deletion of data at the end of the pandemic. From the EU's approach in responding to the Covid-19 epidemic, it can be concluded that the EU confirms the necessity and legality of surveillance under the union laws, GDPR, and national laws. The EU regularly monitors the legal concerns of contact tracing technologies with special reference to privacy and personal data protection principles. Hence, the Commission has developed related guidelines and recommendations. The EU requires that any related mechanism and domestic legislation should be in line with EU laws, constitutional principles, and international law to avoid any unnecessary limitations on fundamental rights. All the documents accentuate the limited retention period, deletion of collected data at the end of the epidemic, notifying individuals on the purposes of data collection by plain language, processing for specific purposes by authorized agencies, prohibition of data sharing with third parties, the necessity of data protection impact assessment, privacy by design and privacy by default mechanisms.

3.4   OECD Guidelines

The Organization for Economic Co-operation and Development (OECD) published a document called *Tracking and tracing COVID: Protecting privacy and data while using*

Table 3. European Union Standards on Contact Tracing Data.

No	Document	Requirements
1	Guidance on Apps supporting the fight against COVID 19 pandemic in relation to data protection	Specific & precise purpose(s); not to bundle different functionalities; not using the gathered data to fight pandemic for other purposes; purposes like scientific research and statistics must be clearly stated under the purposes
2	European Data Protection Board recommendations	Obligation for data protection & respect of human rights, voluntary use of contact-tracing technologies; data protec-tion impact assessments; privacy by design; privacy by default mechanisms; deleting or anonymizing of data at the end
3	Statement on the processing of personal data in the context of the COVID-19 outbreak	Clear, specific, and explicit purposes; lawful processing (under union law and national laws); retention period; Notifying individuals in clear and plain language by easily accessible means; prohibition of data sharing with unauthorized parties; processing of special categories of personal data by competent public authorities and employers under Articles 6, 9, and recital 46 of the GDPR;
4	EU toolbox for mobile applications for contact tracing and warning	People's freedom; single market; fundamental rights; right to privacy; protection of personal data
5	Directive 2011/24/EU of the European Parliament and of the Council of 9 March 2011 on the application of patients' rights in cross-border healthcare	Member states to have clear and specific purposes, processing by authorized personnel, drafting domestic laws for health data processing

apps and biometrics [17]. It explains the current tracking systems and the function of some mobile apps and their risks. It suggests privacy by design and adds that it may involve the use of aggregated, anonymized, or pseudonymous data to maximize privacy protection, or the deletion of data at the end of a pandemic.

It suggests that apps must consider transparency, and robust privacy by design protections, and through open-source projects. Furthermore, it draws the attention that the data must be used for determining purposes and should be deleted after the crisis ends. Before that, the OECD issued a statement *Ensuring data privacy as we battle COVID-19* which also emphasizes the potential risks of some current technologies which may violate the right to privacy and fundamental rights [16]. It points to the lack of transparency and public consultation as the causes of privacy violations by these technologies. While it recognizes the value of data in combating Covid-19, it warns that the way some governments are responding to the pandemic calls for "novel data governance and privacy challenges".

4   Concluding Remarks

When the Minister was not aware of the fact as he said "Frankly, I had not thought of the CPC when I spoke earlier" and since the SNDGO after more than 9 months affirms that "We acknowledge our error in not stating that data from TraceTogether is not exempt from the Criminal Procedure Code" (Singapore Ministry of Communications and Information, 2019) individuals did not have any idea that their data will be shared as well. It can be concluded that Singapore's government's approach was against the right to privacy and personal data protection principles since the App data was for the specific purpose of fighting Covid-19 for a temporary period, considering the lack of primary notification for sharing data for a different purpose. However, the government justifies its decision based on the CPC, the exclusion of public agencies from the application of PDPA, and Covid-19 (Temporary Measures) (Amendment) Act 2021. No one denies the importance of citizens' security, public safety, and justice, but at the same time, data protection laws and individuals' privacy must be observed. The pandemic raises both old and new challenges to privacy and has prompted a flurry of expert reports and statements to guide policymakers and legislators as they consider measures to protect public health. A joint statement of experts from the UN, the Inter-American Commission for Human Rights (IACHR), and the Organization for Security and Co-operation in Europe (OSCE) has cautioned against the invasions of privacy in the name of fighting the COVID-19 pandemic. They suggested limited use of tracking tools, for purposes and time.

One can conclude that while contact tracing technologies are necessary to combat covid-19, they involve high risks. The evaluated instruments emphasize that fundamental rights, privacy, and personal data protection are the core concern to be respected by the surveillance systems. The use of these technologies is voluntary by users. International guidelines draw the attention that the collected data must be processed by an authorized agency for a clear, specific, and limited purpose which is fighting against the pandemic. Recommendations elucidate that data storage is allowed for a limited period and data must be deleted from the main server after the epidemic ends since the systems only apply under emergency conditions. Guidelines warn that sharing of data with third parties is strictly prohibited. The concerns are mostly about sharing data with government agencies which may be used against individuals' freedom and privacy. These recommendations and guidelines collectively require the following measures and principles to be observed by both private and public while using surveillance technologies: security safeguards, proportionality, openness, transparency, accountability, risk management, data protection impact assessment, privacy by design, and privacy by default, public consultation, and consulting with the privacy enforcement authorities. However, the Singapore government's approach was against related international guidelines, the right to privacy, and principles of personal data protection law, and may affect the individuals' trust in the future. As a general and universally accepted principle in all data protection laws, the specific and limited purpose(s) of data collection must be specified at the beginning of data collection. The purpose must be notified to the individuals clearly and in plain language. The data user cannot use the collected data for unrelated purposes or change or add any other purposes after the data is already collected. Sharing of data for other unconnected purposes with third parties is strictly prohibited and against international rules, norms, and practices. In this instance, can the law right the wrong?

Acknowledgment. This research work has been funded by the Permodalan Nasional Berhad (PNB), Project KURSI YAYASAN TUN ISMAIL ALI – UM.0000054/KWG.DM. The authors would like to thank the PNB Malaysia for the financial support.

References

1. Alibeigi, A., Munir, A.B., Ershadulkarim, M.D., Asemi, A.: Towards standard information privacy, innovations of the new general data protection regulation. Libr. Philos. Pract. (e-journal) **2840** (2019)
2. Alibeigi, A., Munir, A.B.: Malaysian personal data protection act, a mysterious application. Univ. Bologna Law Rev. **5**(2), 362–374 (2020). https://doi.org/10.6092/issn.2531-6133/124 412019
3. Bachelet, M.: Coronavirus: human rights need to be front and centre in response, says Bachelet (n.d.). https://www.ohchr.org/EN/NewsEvents/Pages/DisplayNews.aspx?NewsID= 25668. Accessed 27 Feb 2021
4. Buchanan, W., Imran, M., Pagliari, C., Pell, J., Rimpiläinen, S.: Use of participatory apps in contact tracing – options and implications for public health, privacy and trust. Digital Health and Care Institute, University of Strathclyde, Glasgow, Report (2020)
5. Chee, K.: Bill limiting police use of TraceTogether data to serious crimes passed, The Straits Times (2021). https://www.straitstimes.com/singapore/politics/bill-limiting-use-of-tracetoge ther-for-serious-crimes-passed-with-govt-assurances. Accessed 19 Feb 2021
6. Chee, K., Amp Yuen-C, T.: S'pore govt to pass law to ensure TraceTogether data can be used only for serious crimes (2021). https://www.straitstimes.com/singapore/legislation-to-be-passed-to-ensure-tracetogether-data-can-only-be-used-for-serious-crimes#:~:text= SINGAPORE%20%2D%20A%20law%20will%20be,including%20murder%2C%20terr orism%20and%20rape. Accessed 13 Jan 2021
7. Cho, H., Ippolito, D., Yu, Y.W.: Contact tracing mobile apps for COVID-19: Privacy considerations and related trade-offs. arXiv preprint arXiv:2003.11511 (2020)
8. Directive 2011/24/EU of the European Parliament and of the Council of 9 March 2011 on the application of patients' rights in cross- border healthcare (OJ L 88, 4.4.2011, p. 45). https:// eur-lex.europa.eu/LexUriServ/LexUriServ.do?uri=OJ:L:2011:088:0045:0065:en:PDF
9. EDPB: Statement on the processing of personal data in the context of the COVID-19 outbreak (2020). https://edpb.europa.eu/our-work-tools/our-documents/statements/statem ent-processing-personal-data-context-covid-19-outbreak_en. Accessed 22 Feb 2021
10. European Cluster Collaboration Platform: EU toolbox for mobile applications for contact tracing and warning (2020). https://ec.europa.eu/health/sites/default/files/ehealth/docs/covid-19_apps_en.pdf. Accessed 27 Feb 2021
11. European Commission: Commission Recommendation (EU) 2020/518 of 8 April 2020 on a common Union toolbox for the use of technology and data to combat and exit from the COVID-19 crisis, in particular concerning mobile applications and the use of anonymised mobility data", Publications Office of the EU (2020). https://www.scribbr.com/apa-citation-generator/new/webpage/. Accessed 11 Feb 2021

12. European Data Protection Board: Statement by the EDPB Chair on the processing of personal data in the context of the COVID-19 outbreak (2020). https://edpb.europa.eu/news/news/2020/statement-edpb-chair-processing-personal-data-context-covid-19-outbreak_en. Accessed 8 Feb 2021

13. Guidance on Apps supporting the fight against COVID 19 pandemic in relation to data protection. EUR, 17 April 2020. https://eur-lex.europa.eu/legal-content/EN/TXT/?uri=CELEX%3A52020XC0417%2808%29

14. Lapolla, P., Lee, R.: Privacy versus safety in contact-tracing apps for coronavirus disease 2019. Digital Health **6**, 1–2 (2020)

15. McCall, M.K., Skutsch, M., Honey-Roses, J.: Surveillance in the COVID-19 normal - tracking, tracing and snooping: trade-offs in safety and autonomy in the e-city. Int. J. E-Plann. Res. **10**(2), 27–44 (2021)

16. OECD: Ensuring data privacy as we battle COVID-19 (2020). https://www.oecd.org/coronavirus/policy-responses/ensuring-data-privacy-as-we-battle-covid-19-36c2f31e/. Accessed 9 Mar 2021

17. OECD: Tracking and tracing COVID: protecting privacy and data while using apps and biometrics (2020). https://www.oecd.org/coronavirus/policy-responses/tracking-and-tracing-covid-protecting-privacy-and-data-while-using-apps-and-biometrics-8f394636/. Accessed 10 Mar 2021

18. Pierson, D.: Singapore says its contact-tracing data can be used for criminal investigations (2021). https://www.latimes.com/world-nation/story/2021-01-05/singapore-coronavirus-contact-trace-data-criminal-investigations. Accessed 14 Jan 2021

19. Pierucci and Walter: Council of Europe, Joint Statement on Digital Contact Tracing, by Alessandra Pierucci, Chair of the Committee of Convention 108 and Jean-Philippe Walter, Data Protection Commissioner of the Council of Europe. https://rm.coe.int/covid19-joint-statement-28-april/16809e3fd7. Accessed 14 Sept 2021

20. Singapore Ministry of Communications and Information: MCI's response to PQ on public agencies' exemption from PDPA (2019). https://www.mci.gov.sg/search-results#q=MCI's%20response%20to%20PQ%20on%20public%20agencies'%20exemption%20from%20PDPA. Accessed 3 Mar 2021

21. Singapore PDPC Office: Advisories on Collection of Personal Data for COVID-19 Contact Tracing and Use of SafeEntry (2020). https://www.pdpc.gov.sg/Help-and-Resources/2020/03/Advisory-on-Collection-of-Personal-Data-for-COVID-19-Contact-Tracing. Accessed 16 Jan 2021

22. Smart Nation Singapore: Clarification on the Usage of TraceTogether Data by Dr Vivian Balakrishnan (2021). https://www.smartnation.gov.sg/whats-new/speeches/clarification-on-the-usage-of-tracetogether-data-by-dr-vivian-balakrishnan. Accessed 19 Feb 2021

23. Smart Nation Singapore: Upcoming Legislative Provisions for Usage of Data from Digital Contact Tracing Solutions (2021). https://www.smartnation.gov.sg/whats-new/press-releases/upcoming-legislative-provisions-for-usage-of-data-from-digital-contact-tracing-solutions. Accessed 1 Feb 2021

24. Sun, D.: TraceTogether data was accessed in May 2020 for Punggol Fields murder investigations (2021). https://www.straitstimes.com/singapore/politics/tracetogether-data-was-accessed-in-may-2020-for-punggol-fields-murder. Accessed 19 Feb 2021

25. UN Privacy Policy Group: Joint Statement on Data Protection and Privacy in the COVID-19 Response (2020). https://www.who.int/news/item/19-11-2020-joint-statement-on-data-protection-and-privacy-in-the-covid-19-response. Accessed 2 Mar 2021

26. UN Special Rapporteur: United Nations General Assembly, Report of the of the Special Rapporteur on the promotion and protection of the right to freedom of opinion and expression, David Kaye, on disease pandemics and the freedom of opinion and expression, A/HRC/44/49, 23 (2020). https://undocs.org/A/HRC/44/49. Accessed 28 Mar 2021

27. United Nations Secretary General: COVID-19 and Human Rights: We are all in this together (2020). https://unsdg.un.org/resources/covid-19-and-human-rights-we-are-all-tog ether. Accessed 13 Feb 2021

28. WHO Western Pacific Region: World Health Organisation, Considerations for strengthening legal frameworks for digital contact tracing and quarantine tools for COVID-19, Interim Guidance, 15 June 2021

29. World Health Organization: WHO guidelines on ethical issues in public health surveillance (2017). https://www.who.int/ethics/publications/public-health-surveillance/en/. Accessed 5 Sept 2021

30. World Health Organization: Contact tracing in the context of COVID-19: interim guidance, 10 May 2020. https://apps.who.int/iris/handle/10665/332049. License: CC BY-NC-SA 3.0 IGO. Accessed 22 Mar 2021

31. World Health Organization: Ethical considerations to guide the use of digital proximity track-ing technologies for COVID-19 contact tracing (2020). https://www.who.int/publications/i/item/WHO-2019-nCoV-Ethics_Contact_tracing_apps-2020.1. Accessed 11 Sept 2021

32. World Health Organization: Contact tracing in the context of COVID-19: interim guidance, 1 February 2021. http://www.WHO/2019-nCoV/Contact_tracing_and_quarantine/2022.1. Accessed 23 Mar 2021

Sentiment Analysis Application in E-Commerce: Current Models and Future Directions

Huang Huang[(✉)] [iD], Adeleh Asemi [iD], and Mumtaz Begum Mustafa [iD]

Malaya University, Kuala Lumpur, Malaysia
erichuangemail@163.com

Abstract. Sentiment analysis (SA), which is also known as opinion mining, is an increasingly popular practical application of Natural Language Processing (NLP). SA is especially useful in e-commerce fields, where comments and reviews often contain a wealth of valuable business information that has great research value. This study aims to investigate three related aspects of SA in e-commerce: the methods used to address the SA problem in this domain, the most commonly used e-commerce platforms where researchers get data from, and the future direction of research in this area. To achieve this goal, we reviewed 15 papers that covered many machine learning models and deep learning models. In the results and discussion section, we suggest several future directions that can improve the current SA models in this review. By addressing the limitations of existing SA models and exploring new approaches, we believe that future research in this area will lead to more accurate and effective sentiment analysis tools that can benefit both businesses and consumers.

Keywords: Sentiment analysis (SA) · e-commerce · Natural language processing · Machine learning · Deep learning · Opinion mining

1 Introduction

The rapid development of smart phone, Internet users are changing from traditional information receivers to information publishers, it has revealed that e-commerce websites offer a significant volume of valuable information that surpasses the cognitive processing abilities of humans [3]. Zhang mentioned microblogs comprise of complex and copious sentiments that depict the user's perspectives or viewpoints regarding a particular subject [15].

Sentiment analysis is crucial in e-commerce, helping businesses analyze customer feedback, identify trends, and summarize popular platforms. This paper provides a summary of commonly used platforms and an overview of sentiment analysis techniques in e-commerce and future directions for researchers. In this study, Sect. 2 outlines the methodology for reviewing sentiment analysis techniques in e-commerce. Section 3 presents the results and discussions for each research question. Lastly, Sect. 4 summarizes the conclusions derived from this review study.

© The Author(s), under exclusive license to Springer Nature Switzerland AG 2023
A. Kö et al. (Eds.): EGOVIS 2023, LNCS 14149, pp. 67–72, 2023.
https://doi.org/10.1007/978-3-031-39841-4_5

2   Review Method

Research Questions Raised in This Review

Here are the research questions to be addressed in this research. RQ1: What are the methods used to solve sentiment analysis problem in e-commerce? RQ2: What are the types of the e-commerce platform used to apply sentiment analysis? RQ3: What are the future directions/issue for sentiment analysis in e-commerce?

Search Methodology

The key words are: sentiment analysis and e-commerce. We applied inclusion criteria to filter and include relevant studies in our analysis: Search domain: AI, Computer Science or Information Technology; Publication type: Journals; Language: English; We used exclusion criteria to remove irrelevant papers from consideration: Papers that do not specifically focus on sentiment analysis and e-commerce; Papers with sentiment analysis as a supplementary subject; Paper without detailed experiments;

Search Results and Analysis

This scholarly paper comprehensively examines related academic papers published between 2018 and 2023. Due to the rapid advancement of sentiment analysis technologies in the field of e-commerce, our focus is limited to reviewing papers published within the past 5 years. After conducting a keyword search, we identified 67 papers. Through manual analysis of titles and abstracts, 33 irrelevant papers were removed, resulting in a final selection of 34 papers. We then read the full papers and applied the exclusion criteria, which resulted in the rejection of 19 papers and left us with 15 relevant papers. Ultimately, the final number of papers selected for the review study was 15. Regarding the origins of the selected experimental papers, it can be observed that 4 of them, which accounts for 26.7%, were obtained from the IEEE database. Additionally, 10 papers were retrieved from Google Scholar, while another one originated from the Science Direct database. Out of the 15 papers analyzed, 7 (46.7%) solely employed machine learning techniques, while 8 (53.3%) exclusively utilized deep learning methods. These findings indicate that machine learning and deep learning techniques hold equal weight in the realm of sentiment analysis.

3   Results and Discussion

3.1   RQ1: What Is the Method Used to Solve Sentiment Analysis Problem in E-Commerce?

Sentiment analysis is vital in natural language processing, with two commonly used approaches: machine learning methods and deep learning methods.

Machine Learning Based Methods

In sentiment analysis classification task, many machine learning algorithms are applied. Logistic Regression (LG) [2,7], Naïve Bayes (NB) [2,4,7,14] and Support Vector Machine (SVM) [4,7] are the most welcome algorithms.

Jagdale's study compares NB and SVM in various comment datasets, including camera, laptops, mobile phones, tablets, TVs, and video surveillance. The comparison is based on different criteria such as accuracy, precision, and F-score. Across all datasets and criteria, NB consistently outperforms SVM [4]. For example, the accuracy measurements for NB and SVM classifier for the dataset. NB classifier got 98.17% accuracy for Camera reviews and Support Vector Machine got 93.54% accuracy for Camera reviews.

In Elmurngi's study, they provide comparison of four supervised machine learning algorithms: Naïve Bayes (NB), Decision Tree (DT-J48), Logistic Regression (LR) and Support Vector Machine (SVM) for sentiment classification using three datasets of reviews, including Clothing, Shoes and Jewelry reviews, Baby reviews as well as Pet Supplies reviews. To assess the performance of sentiment classification, their study has employed accuracy, precision, and recall as performance measures. Their experimental results demonstrate that the Logistic Regression (LR) algorithm outperforms the other three classifiers, not only in text classification but also in detecting unfair reviews. LR exhibits the highest accuracy, making it the preferred classifier in their work [7].

Deep Learning Based Methods

In recent studies, CNN (Convolutional Neural Network) [3,5,11], RNN (Recurrent Neural Network) [10], LSTM(Long Short-Term Memory Networks) [6,10] and BERT [3] are the most frequently used.

Wei Hong and his team focused on online review information and used a convolutional neural network(CNN) text mining model for its analysis [5]. CNNs' self-learning capabilities were utilized to train word vectors, mitigating issues related to data sparsity. In their study, the text classification model based on convolutional neural network (CNN) was constructed using a single convolution layer and pooling layer during training. CNN has gained significant popularity in the field of computer vision. Its application to NLP represents a courageous and innovative endeavor. Prior to the introduction of the GPT-4 model, BERT held the position as the state-of-the-art solution for NLP problems. Alhassan Mabrouk and his team found that the opinion summarization task witnessed impressive performance from the RNN-based BERT model across various LC5 benchmark datasets [3].

Table 1. Methods, Platforms and future directions of papers.

Ref No.	Method	Platform	Future Direction
[1]	PTSM (Machine Learning)	gross box office	Different language models
[2]	SVC (Machine Learning)	Amazon	multi-gram feature sets
[3]	Bert, CNN, RNN (Deep Learning)	Amazon	Other domains
[4]	Lexicon based (machine Learning)	Amazon	Aspect level SA
[5]	CNN (Deep Learning)	Alibaba and JD	Other domains
[4]	Lexcion based (Machine Learning)	Amazon	Improve lexicon
[6]	LSTM (Deep Learning)	Traveloka	NA
[7]	Machine Learning	Amazon	Select different features
[8]	CNN-BiLSTM (Deep Learning)	SemEval 2014 twitter dataset	NA
[9]	Deep Learning	TripAdvisor	Different features selection
[10]	RNN, LSTM (Deep Learning)	E-Commerce Reviews	hyper-parameter tuning
[11]	CNN, BiGRU (Deep Learning)	dangdang.com	sentiment refinement
[12]	Deep Learning	Amazon	NA
[13]	Machine learning	Amazon	automate data labeling
[14]	Machine Learning	Amazon	Other domains data

3.2 RQ2: What Is the Type of the E-Commerce Platform Used to Apply Sentiment Analysis?

Based on Table 1, it is evident that Amazon [2–4, 7, 12–14] is the primary platform from which researchers collect data, with 9 out of 15 papers utilizing it. Amazon, being the largest e-commerce platform globally, understandably serves as the primary data source for many research papers. However, it is important to note that Amazon's widespread usage is primarily limited to English-speaking countries. As a result, the models developed using Amazon's data are more applicable to English language tasks. On the other hand, for languages such as Chinese and Arabic, the applicability of these models is relatively lower.

Social media platforms such as Twitter [8] are one of data sources, with 1 out of 15 papers using them. Twitter differs from online e-commerce platforms in the sense that people share opinions more casually. Information is often contained within a single tweet, making it more challenging for models to analyze compared to professional e-commerce platforms like Amazon. Besides the previously mentioned platforms, there exist numerous other online platforms where users can express their opinions and share their experiences regarding their travels such as dangdang.com [11], TripAdvisor [9] and Traveloka [6]. In other industries, Data comes from different platforms but all of them are in service industries platforms including restaurant and reviews from gross box office [1], The Alibaba Group and JD [5].

3.3   RQ3: Future Directions/Issue for Sentiment Analysis in E-Commerce?

More Universal Models in New Domains and Languages

In most of implemented models, the data source is most in English, therefore, those models can only performs better at predicting English inputs. It is still a promising direction to get sentiment reviews in different languages other than English to conduct cross-cultural studies in this field [1]. For one specific machine learning or deep learning model, in most situations, it can only performs well in one specific domain [3]. An universal model which can be applied across domains still has a long way to go.

Aspect-Level Sentiment Analysis Models

Aspect-Level sentiment analysis is still a promising future direction [4]. In recent researches, it has been observed that there is a conspicuous paucity of aspect-level sentiment analysis language models. This implies that there is a pressing need for the development of more advanced machine learning or deep learning models that can precisely extract and analyze aspects and their corresponding sentiments.

Lexicon Based Models Introducing External Knowledge

Incorporating external knowledge resources in the form of lexicons and ontologies is a promising approach to enhancing the performance of natural language processing models [4], providing greater accuracy, flexibility, and adaptability for a wide range of applications.

4   Conclusion

This paper summarizes recent studies on sentiment analysis in e-commerce, categorizing the models based on approaches and e-commerce platforms. It also highlights the significance of further research in this domain. Sentiment analysis applied in e-commerce still heavily rely on machine learning and deep learning methods, each with their own advantages and disadvantages. LG, NB, and SVM are the commonly employed machine learning methods. The application of CNN in NLP represents an innovative approach, while BERT continues to be the state-of-the-art model in the field before GPT-4. The primary data sources for sentiment analysis, such as Amazon, Twitter, continue to be crucial. Compared to Amazon, comments on Twitter are generally more casual, which poses additional challenges in the modeling process. To advance sentiment analysis in e-commerce, future studies could focus on developing more universal models for new domains and languages, aspect-level sentiment analysis models and lexicon-based models introducing external knowledge. It is hoped that more attention will be given to these areas.

References

1. Huang, L., Dou, Z., Hu, Y., Huang, R.: Textual analysis for online reviews: a polymerization topic sentiment model. IEEE Access **7**, 91940–91945 (2019)
2. Ahmed, H.M., Javed Awan, M., Khan, N.S., Yasin, A., Faisal Shehzad, H.M.: Sentiment analysis of online food reviews using big data analytics. Elementary Educ. Online **20**(2), 827–836 (2021)
3. Mabrouk, A., Redondo, R.P.D., Kayed, M.: SEOpinion: summarization and exploration of opinion from e-commerce websites. Sensors **21**(2), 636 (2021)
4. Jagdale, R.S., Shirsat, V.S., Deshmukh, S.N.: Sentiment analysis on product reviews using machine learning techniques. In: Mallick, P.K., Balas, V.E., Bhoi, A.K., Zobaa, A.F. (eds.) Cognitive Informatics and Soft Computing. AISC, vol. 768, pp. 639–647. Springer, Singapore (2019). https://doi.org/10.1007/978-981-13-0617-4_61
5. Hong, W., Zheng, C., Wu, L., Pu, X.: Analyzing the relationship between consumer satisfaction and fresh e-commerce logistics service using text mining techniques. Sustainability **11**(13), 3570 (2019)
6. Muhammad, P.F., Kusumaningrum, R., Wibowo, A.: Sentiment analysis using Word2vec and long short-term memory (LSTM) for Indonesian hotel reviews. Procedia Comput. Sci. **179**, 728–735 (2021)
7. Elmurngi, E.I., Gherbi, A.: Unfair reviews detection on Amazon reviews using sentiment analysis with supervised learning techniques. J. Comput. Sci. **14**(5), 714–726 (2018)
8. Meng, W., Wei, Y., Liu, P., Zhu, Z., Yin, H.: Aspect based sentiment analysis with feature enhanced attention CNN-BiLSTM. IEEE Access **7**, 167240–167249 (2019)
9. Liang, X., Liu, P., Wang, Z.: Hotel selection utilizing online reviews: a novel decision support model based on sentiment analysis and DL-VIKOR method. Technol. Econ. Dev. Econ. **25**(6), 1139–1161 (2019)
10. Agarap, A.F.: Statistical analysis on E-commerce reviews, with sentiment classification using bidirectional recurrent neural network (RNN). arXiv preprint arXiv:1805.03687 (2018)
11. Yang, L., Li, Y., Wang, J., Sherratt, R.S.: Sentiment analysis for E-commerce product reviews in Chinese based on sentiment lexicon and deep learning. IEEE Access **8**, 23522–23530 (2020)
12. Shrestha, N., Nasoz, F.: Deep learning sentiment analysis of amazon.com reviews and ratings. arXiv preprint arXiv:1904.04096 (2019)
13. Haque, T.U., Saber, N.N., Shah, F.M.: Sentiment analysis on large scale Amazon product reviews. In: 2018 IEEE International Conference on Innovative Research and Development (ICIRD), pp. 1–6. IEEE, May 2018
14. Sari, P.K., Alamsyah, A., Wibowo, S.: Measuring e-Commerce service quality from online customer review using sentiment analysis. J. Phys. Conf. Ser. **971**(1), 012053 (2018)
15. Zhang, S., Wei, Z., Wang, Y., Liao, T.: Sentiment analysis of Chinese micro-blog text based on extended sentiment dictionary. Future Gener. Comput. Syst. **81**, 395–403 (2018)

Artificial Intelligence

Evolving Justice Sector: An Innovative Proposal for Introducing AI-Based Techniques in Court Offices

Flora Amato[1], Simona Fioretto[1], Eugenio Forgillo[2], Elio Masciari[1], Nicola Mazzocca[1], Sabrina Merola[1], and Enea Vincenzo Napolitano[1]

[1] Department of Electrical and Information Technology Engineering, University of Naples Federico II, Naples, Italy
`eneavincenzo.napolitano@unina.it`
[2] Court of Appeal of Naples, Naples, Italy

Abstract. The digital transformation of public administrations (PA) plays a central role in the social and economic growth of a country. Currently, the application of Information Technology in government services, namely e-government, is still in its infancy and many applications are new born. The use of Artificial Intelligence (AI) could help improve the digital transformation of the PA. Although AI is a hot topic that is already used in everyday life and is constantly evolving, when it comes to the public sector, and in particular the justice sector, it is always difficult to think and recognise AI applications that actually help citizens and society. The main goal of this paper is to explore AI applications in the public domain to show how a real application in the justice sector could be helpful. Then, some results of real applications in the Court of Appeal of Naples are discussed and some future directions are proposed.

Keywords: E-Government · Public Administration · Artificial Intelligence

1 Introduction

Since 2014, the European Commission has annually assessed the digitalization of member states through the Digital Economy and Society Index (DESI) [7]. The DESI index provides information on the digitalisation of European countries in different areas by analysing data from the European Statistical Office (EUROSTAT). The full online delivery of public services is one of the targets proposed by the Commission to be achieved by 2030. In this context, many studies considered information technology tools as well as artificial intelligence to achieve the goals and objectives set by the European Commission. Despite its use in different areas today, the application of Artificial Intelligence to improve the online availability of public services, such as e-government, is a fairly new topic [1]. The justice sector is the most recent area of application of these innovations and techniques. In fact, as widely discussed and shared by [21], law is very different from other fields such as medicine and finance. The actual courtroom process still looks a lot like it did in 1980. Therefore, there is a need for a proper digital transformation as citizens, companies and organisations are willing to change the justice system, which is extremely slow. Disputants are no longer willing to pay large retainers and be billed for expensive lawyer time to resolve their cases over a long period of time [21]. In fact, of all the areas, justice, with its long delays, has the greatest impact on society and the economy, and there are certainly several challenges and opportunities to be identified and innovations to be introduced. In this paper we will discuss our proposal based on the results of an important research project in which the Naples Court Office has participated.

2   Related Work

AI applications in the public sector differ from those in industry due to the specific characteristics of the public sector. In fact, the public domain is peculiar and is subject to different rules and constraints that influence its daily operation and therefore make it unique. The aim of this section is to analyse, through a brief review of the literature, the applications of AI in the PA, with a particular focus on the judiciary, addressing the advantages they bring, the limitations they are subject to, and the desired future developments. First, we will review the applications of AI for PA and then we will focus on the specific environment of the Italian judicial bodies in order to provide a complete picture of the current level of digitalisation.

The increasing trend of AI applications did not turn out to be a parallel increasing trend in the literature observations [25]. Through a literature review, the authors organise the research results into 5 categories:

1. AI public service: deals with AI applications that focus on improving the time and quality of services provided;
2. Working and social environment affected by AI: looks at the impact of AI on the social environment;
3. Public order and law related to AI: these two areas can be supported by AI applications for predictive models to minimise damage and casualties from natural disasters and to support surveillance by government agencies;
4. AI ethics: assessing the benefits and threats to society;
5. AI government policy: assessing laws and policies that AI might consider and how they might adapt to AI applications.

The first category is of particular interest as it relates to AI applications that can support government services and improve the efficiency of public administration. The authors offer a structured table containing all relevant applications of AI in PA, divided into categories. Based on our research goal of improving online government services through AI integration, we decided to analyse the platforms proposed in [6,26]. The latter papers refer to AI applications that improve government services through workflow standardisation and automation [19], virtual agents, and automatic task assignment. In [1], a proposal for an intelligent e-government platform architecture is presented that supports the development and implementation of e-government AI applications. In particular, the functional layer of the proposed architecture aims to fulfil citizen requests in four phases with the help of virtual agents and AI techniques.

From this initial research analysis, it is possible to state that emerging AI applications supporting e-government services have the ultimate goal of streamlining the process of satisfying citizen requests, doing so with high quality and as much trustworthiness as possible.

Another line of research in the legal field related to AI applications is the search for charities in legal documents through the technique of Named Entity Recognition (NER). For example, in [10], the NER technique is used to search for all legal entities within US legal documents, such as: judges, jurisdictions and courts. The system described in [10] is a system based on three main tools:

1. Search tables;
2. Context rules;
3. Statistical models.

In [11] the authors propose the realisation of a system dedicated to the support of the lawyer's acumen and the automatic creation of the argument in relation to the case.

With regard to the European context, and in particular the Italian one, it can be said that the system is evolving, certainly driven by the pandemic. Although the sector is growing, it is still very much linked to the traditions established within the procedures, which are mostly based on manual practices, which implies a slow development. At the moment, Italian judicial procedures, both criminal and civil, are not very digitised. There are no automatic tools that allow to manage and control a process or a trial during its development. Over the years, tools have been provided to support some procedures rather than others, covering only some parts and tasks of the process. In addition, courtrooms are not equipped with hardware and software tools, which means that the digital debate of a trial is not guaranteed. The legal domain has proven to be unfriendly to AI applications, which may be due to the specificity and uniqueness of each country's legal system. As each country has its own rules and language, it is difficult to define a common standard. In addition, an important emerging challenge is that the majority of studies on the subject use English legal documents, which hinders the growth of legal digitisation in countries where different languages are spoken.

3 Case Study

The solutions proposed in this paper are the result of a field analysis in the judicial offices; in particular, the work was carried out in different time periods from September 2022 to February 2023, mainly in the Court of Appeal of Naples. In order to identify the working areas and the study methods and techniques to be used, it is necessary to understand the working scenarios of the offices studied.

The preliminary phase aims at analysing the internal processes of the judicial offices of Naples in order to identify the critical points where the use of AI can bring significant improvements. In this context, we are conducting a series of interviews and focus groups (webinars) with judges, lawyers, judicial officials and other actors involved in the judicial process. We also analyse information flows, IT systems and available resources, and use all the techniques necessary to obtain information on the efficiency of offices.

3.1   Analysis

The analysis and field mapping phase is the most critical step in the entire digitalization process. In fact, it is constantly evolving during the project. Given the specific nature of the research, it was not easy to identify the requirements, areas for improvement, challenges and then the software specifications. To this end, the analysis was carried out using the Agile methodology. The choice of the Agile methodology is based on its characteristics as highlighted in [15]. More specifically, agile software development should start with simple and predictable approximations of the final requirements and then continue to increase the detail of these requirements throughout the project lifecycle. This incremental refinement continuously improves the design, coding and testing at all stages of production activity.

Following the requirements of the agile methodology, we carried out two iterations:

1. **First Iteration**: The aim of this phase is to start the project. We planned several meetings with the client, who in our case were legal actors. After the meetings we did a big brainstorming with the working group. Then we set the objectives to be achieved after this first phase. Then we designed and developed software and solutions based on the requirements. Finally, the solutions were presented to the clients for feedback to be incorporated in the second iteration;
2. **Second Iteration**: From the first iteration we received some feedback that helped us to set up the second iteration. So we decided to gather new information about the software requirements through structured interviews based on the previous feedback and through field analysis with the actors. This new phase also needed to explore existing data, platforms and software solutions used in the office.

3.2   Data Structure

During the analysis phase, we identified a number of data retrieval challenges. For example, the existing data contained a lot of sensitive information due to the characteristics of the field, which made it difficult the direct access. In addition, we did not find a centralised database from which to retrieve information.

To overcome this problem, we collect data from different sources. We select different data for the two solutions:

1. For classification and anonymisation we use
   - *Legal citation*[1] for the classification tool;
   - *Judgments of the Court of Cassation*[2] for the anonymisation tool.
2. For the management of human resources and the allocation of workload, we use data extracted from the Court of Appeal of Naples, which can be summarised as follows:
   - Judges in charge: Personal data of the judges, with current and historical positions;
   - Assigned staff: Personal details of secretarial and administrative staff, with related functions;
   - Sections: Details of the sections into which the Court of Appeal is divided and their composition;
   - Offices: Physical data and composition of the offices;
   - District offices: Data concerning the offices of the courts of the district, but not of the Court of Naples itself.
3. For the management of the documented flows we use the relative data to the informative flows of the offices, which have been analysed and used to understand the flow of documents.

3.3   Identified Issues

The examination of the data and the working methods of the Offices raises a number of questions and challenges:

1. The possibility of making findings on large quantities of digital documents written in natural language is a relevant activity that could have a great impact on the extraction of information from PA. In the offices analysed, a distinction is made between formal written documents and digital documents. In fact, the formal written document, called "act", is considered the only one with official value for official purposes. The digitisation of documents is only considered for administrative purposes such as classification and information extraction. This is linked to the need to have both document formats, but since the "act" is the only official one, the digital format is often considered of secondary importance, which sometimes leads to the absence of the digital document.
2. Human Resources Management (HRM) is a fundamental aspect within the judicial offices. These offices manage a staff consisting of a mix of senior legal functions. The analyses carried out show that it is difficult to manage staff time in terms of
   - Non-scheduled working time;

---

[1] The dataset is available here: https://www.kaggle.com/datasets/shivamb/legal-citation-text-classification.

[2] The dataset is available here: https://www.italgiure.giustizia.it/sncass/.

- Holidays;
- Standard tasks;
- Occasional additional tasks.

In fact, in these offices there are no information systems that can manage the work activities of the staff. The above activity is done manually, mixing information coming from different sources, and consequently it is affected by human errors and, being inflexible, it requires time to manage setbacks. In addition, the allocation of tasks is influenced by HRM, as it is not possible to check the availability of staff (monitoring of work records and status of previously allocated tasks) and to find standard criteria for matching users and tasks.

3. The management of document flows is one of the main criticisms. The high volume of documents to be managed, the complexity of judicial proceedings and the need to ensure the security and integrity of information are the main challenges for the proper functioning of judicial processes [5]. Since the documents are related to the preparation of the final case file (which is the one used in court), it is important to check what stage of activity they are in, where they are physically located and who is working on them, in order to achieve full traceability of the document.

After the recognition phase, the intervention areas for improvement are identified. These areas are linked to critical issues such as:

- the lack of traceability of documents;
- the lack of traceability of personnel;
- the overload of staff activities.

The tools proposed in the following sections are our proposals for solving the above problems.

4   Our Proposals

The findings of the survey phase indicate that the efficiency of the Court Office could be improved in four main areas: *smart document classification, personnel monitoring capabilities, process and task management and document management.* Based on the results of the exploration phase, we develop a series of proposals for the implementation of AI-based solutions. Specifically, the first approach relates to an innovative solution that is already operational; the others are projects at an advanced stage of development.

In this section we describe an innovative approach, based on the use of AI, for the optimisation of e-government in the Italian Judiciary. In the following subsections, we provide an overview of the proposed tools, focusing on their functionalities; an in-depth analysis and explanation of each tool is beyond the scope of this paper.

4.1   Classification of Judgments and Anonymization

Text classification and anonymization are based on the application of artificial intelligence techniques. Text classification is a specific task of text mining, which takes a digital text written in natural language and automatically classifies it into one or more defined categories [17]. This task can be useful for the management of the texts of the document flows described above Sect. 3.3.

We decided to solve the classification problem with deep learning solutions, specifically using Transformers. The model chosen is DistilBERT [23] which is literally the distilled and compact version of Bert, which requires less computational resources. In the literature it turned out to have excellent performances on this task, in fact, as reported by [23], the results obtained comparing the pre-addestrated models suggests that although DistilBERT is the compact version of Bert, does not lose consciousness.

The proposed model receives a dossier as input, and automatically classifies it into a defined category which refers to the result of the legal judgment:

- affirmed
- applied
- approved
- cited
- considered
- discussed
- distinguished
- followed
- referred to
- related

The metrics used for the evaluation are the precision, recall and F1 score. The obtained results are showed in Fig. 1.

Another application developed with AI techniques is the anonymization of texts, based on Name Entity Recognition (NER) [18].

This technique aims to identify associations within the text and associate them to a tag [13]. In this context, in order to make the legal judgments anonymous while respecting the privacy of the actors, the previously labelled information was hidden. In our approach, NER is implemented using a rules-based approach. Syntax "rules" and patterns have been defined according to the privacy problem of the actors (Fig. 2).

The advantage of using these two tools is to speed up the work outside the processes. With the first tool we have the possibility to classify the texts in a simple and automatic way, the second tool allows a quick anonymization, which allows to preserve the privacy of the actors involved in the processes and to reduce the difficulties of access to sensitive information.

4.2   Workflow Management

Due to aforementioned reasons in Sect. 3.3, we decided to implement a human resources management system with the aim of supporting and improving the

```
F1 score 0.8839552167956041
Classification Report
                  precision    recall  f1-score   support

       affirmed       1.00      0.99      1.00       370
        applied       0.81      0.79      0.80       361
       approved       1.00      0.73      0.84        11
          cited       0.92      0.95      0.93      3109
     considered       0.66      0.67      0.66       193
      discussed       0.76      0.70      0.73        89
  distinguished       0.83      0.69      0.75        64
       followed       0.82      0.82      0.82       368
    referred to       0.81      0.69      0.74       491
        related       0.97      0.84      0.90        44

       accuracy                          0.89      5100
      macro avg       0.86      0.79      0.82      5100
   weighted avg       0.88      0.89      0.88      5100
```

Fig. 1. Report of evaluation of Model with outcome for each class : precision, recall and F1.

Fig. 2. Example of how a sentence is anonymized by the tool of anonymization.

management of human resources, guaranteeing flexibility, availability and control. The system provides different types of information about the personnel, showing personal data and work data::

- Personal data: refers to both biographical information and a history of previous work experience, relating to seniority and positions occupied;
- Work data: provide an instant picture of the situation in the office, from which it is possible to make views of the workforce and the workers, where the workforce is the side of the staff that is actually working, and the workers are the side of the staff that should be working, but due to external motivation (illness, business trips, etc.) cannot carry out daily activities.

This aspect must be considered because it influences the efficiency of judicial offices which largely depends on the optimization of task allocation [4]. A fair and balanced allocation of tasks among the various members of the judiciary is essential to ensure the effectiveness of the service to citizens. However, the allocation of tasks is a complex task that requires the consideration of many factors and parameters [9]. This problem could be approached by solving an optimisation problem, minimising a function using algorithms that take input parameters on:

- Staff members: including availability, skills and abilities;
- Tasks: type and complexity;
- Workload: volume of work to be handled;

returning as output the best user-task match.

Unfortunately, it is not currently possible to tackle the problem in this way because there are no standard criteria for task allocation. Often, the allocation of tasks is based on subjective decisions made by the supervisor or office manager, which can lead to an unfair allocation of tasks [20]. In fact, we are not able to establish input weights.

To avoid this problem, it is important to establish objective and standard criteria for task allocation. For example, the allocation of tasks can be based on the nature and complexity of the tasks, the volume of work to be handled, and the availability and skills of individual members of staff. This will ensure that the allocation of tasks is based on an objective assessment of the needs of the office and the skills of individual members of staff [22].

Another important factor to be taken into account in workloads assignment is the skills evaluation [12] of individual staff members. Each member of staff has different skills and abilities that need to be taken into account when assigning tasks. Various tools can be used to assess the competencies of individual personnel, such as analysis of past performance, assessment of competencies through specific tests, or self-assessment. Using these tools, it is possible to identify the skills and areas for improvement of individual employees and to assign them the most appropriate tasks based on their skills and abilities.

Finally, it is important to continuously monitor and evaluate the effectiveness of workload assignment. This allows to identify any problems or inefficiencies [8] in the assignment of tasks and to make the necessary adjustments to improve the effectiveness and efficiency of the judicial office.

Monitoring can be carried out by analysing the performance and results of the judiciary, analysing the time spent on different tasks, receiving feedback from staff and from citizens or businesses that interact with the judiciary [3].

Evaluation, on the other hand, can be carried out through the use of performance indicators [14], such as the average response time, the percentage of cases successfully managed or the degree of satisfaction of managers. In this way, it is possible to identify areas for improvement and make the necessary adjustments to ensure a better allocation of workloads and improved efficiency of the legal departments.

Our proposal for workload management efficiency is a software that supports the case manager's decisions. The proposed solution does not change the current flow of information, but simply helps to avoid overloading individual employees or taking advantage of others' free time. Graphs and statistics provide a real-time view of the current status of tasks and the state of the office. In this way, it is possible to refer to an overview dashboard at the time of allocation, before making a decision, allowing a data-driven choice with a lower probability of error.

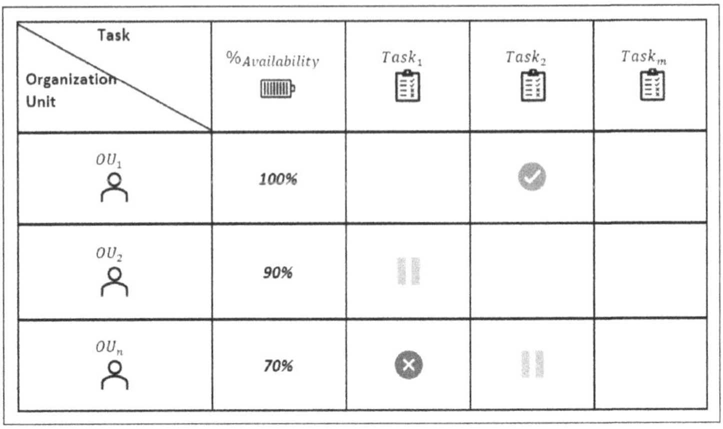

Fig. 3. A schematic representation of the job assignment taking into account the commitment of the individual judge and the history of the assignments.

Future work could lead to the use of optimisation algorithms first, and artificial intelligence later, once a consistent and real database is obtained (Fig. 3).

In conclusion, the allocation of workload in a judicial office is a complex task that requires the use of objective and standard criteria, the assessment of the competencies of individual staff members and the continuous monitoring and evaluation of the effectiveness of the allocation of workload. The adoption of these practices can help to improve the efficiency of the judiciary and ensure a more effective and satisfactory service.

4.3   Document Flow Management

To address the critical issues Sect. 3.3, the proposal in this section is the implementation of a document flow management system based on digital technologies, capable of simplifying and automating document management procedures, specifically improving process efficiency, guaranteeing security and information integrity and real-time monitoring. The system is born as a support to minimise the efforts made in the preparation phase of the file on its path of creation. Knowing in advance the itinerary of each different file, the system can act as a visual tracking map for the same [24].

The aim of the proposed solution is to be able to see the position and the operator who is working on the document. The system will be able to mark each step with the current status and, if necessary, alert the operator to allow the correct transition to the next step. The document flow management system could also involve the adoption of advanced technologies for managing documents in digital format [16].

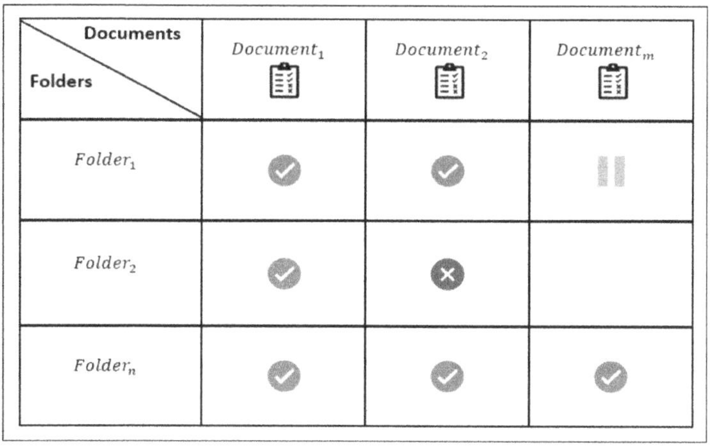

Documents / Folders	$Document_1$	$Document_2$	$Document_m$
$Folder_1$	✓	✓	
$Folder_2$	✓	✗	
$Folder_n$	✓	✓	✓

Fig. 4. A schematic representation of the management of the documents for the monitoring of the state of the single phases of elaboration documents.

In particular, the system is based on the use of document management software, which enables data relating to documents to be collected, processed, stored and managed in an efficient and secure manner. It is important to emphasise that no sensitive data is used internally in this process, but is simply treated as a closed system in order to preserve anonymity and privacy as much as possible (Fig. 4).

Future document management software may include advanced document processing capabilities such as optical document scanning, optical character recognition (OCR), document indexing or code readers for digital codes or QR codes. With these features, the system could be able to capture documents in a matter of seconds, enabling instant localisation, rapid search and retrieval of information. The system also provides for the adoption of advanced security procedures to ensure the integrity and security of documents.

In particular, the document management software could be integrated with the previously introduced anonymisation and classification algorithms to protect documents and guarantee their confidentiality and authenticity.

Finally, the documented case management system could be made available as an integration of the IT systems already used in judicial offices. This integration would make it possible to automate judicial processes and ensure complete traceability of information.

In conclusion, the implementation of a document flow management system is an effective solution to the critical issues faced by judicial offices in relation to document management. Thanks to digital technologies, the system makes it possible to simplify and automate document management procedures, improve process efficiency, guarantee the security and integrity of information, speed up searches, streamline processes and reduce backlogs.

5   Conclusion

Nowadays, e-government is one of the areas that require more attention. A good implementation of information technology techniques supporting government services makes it possible to improve the digitisation of PA, which is one of the objectives of DESI [7]. While some sectors of the PA, such as the health system [2], are already in the process of digital transformation, the legal area requires particular attention, representing one of the sectors with the most difficulties to face. The aim of this paper is to highlight the importance of e-government, with particular reference to the Italian legal context.

After analysing the current challenges in the sector, we carried out analyses and surveys in the field in order to propose technological solutions capable of supporting this process of digital transformation of the Public Administration. Our research highlighted how digital transformation can improve the efficiency, accessibility and transparency of legal services, with the aim of simplifying processes and reducing execution times. The proposed technological solutions can be summarised as follows: a judgement classification system and a document anonymiser, a human resources management tool, an intelligent workload allocation algorithm and a document flow management tool. These proposals can play a crucial role in accelerating the process of digital transformation in the legal field and in ensuring the protection and security of sensitive information.

In conclusion, this study represents a significant and innovative contribution to the digitalisation of the Italian judicial system, as it proposes techniques for analysis, proposes solutions that can represent standards in the approach to the digitalisation of public administrations, and provides interesting bases for future developments and insights on this topic.

Future developments concern the possibility of identifying standard metrics and parameters to carry out the automatic allocation of resources, supporting the 3 proposed tools:

- Human Resource Management Sect. 4.2
- Workload Allocation Sect. 4.2
- Document Flow Management Sect. 4.3

In the future, we intend to analyse the previous Italian proposal and experiences in the same topic and sector. In addition, it is necessary to have practical responses, testing the classification and anonymisation tasks on an Italian dataset and test the proposed tools in the selected offices.

Acknowledgment. Work supported by the project "MOD-UPP" - Macroarea 4 - project PON_MDG_1.4.1_17- PON GOV grant.

We acknowledge financial support from the project PNRR MUR project PE0000013-FAIR.

References

1. Al-Mushayt, O.S.: Automating e-government services with artificial intelligence. IEEE Access **7**, 146821–146829 (2019)

2. Anniciello, A., Fioretto, S., Masciari, E., Napolitano, E.V.: Covid-19 impact on health information technology: the rapid rise of e-health and big data driven innovation of healthcare processes. In: 2022 IEEE International Conference on Bioinformatics and Biomedicine (BIBM), pp. 2759–2764. IEEE (2022)

3. Bedwell, W.L., Salas, E., Funke, G.J., Knott, B.A.: Team workload: a multilevel perspective. Organ. Psychol. Rev. **4**(2), 99–123 (2014)

4. Biber, E.: The importance of resource allocation in administrative law. Admin. L. Rev. **60**, 1 (2008)

5. Chimwani, B.I., Iravo, M.A., Tirimba, O.I.: Factors influencing procurement performance in the Kenyan public sector: case study of the state law office. Int. J. Innov. Appl. Stud. **9**(4), 1626 (2014)

6. Chun, A.H.W.: An AI framework for the automatic assessment of e-government forms. AI Mag. **29**(1), 52–52 (2008)

7. Commission, E.: Digital Economy and Society Index (DESI) 2022. International series of monographs on physics, European Commission, keywords = Digital Economy, Society Index (2022)

8. Cusatelli, C., Giacalone, M.: Evaluation indices of the judicial system and ICT developments in civil procedure. Procedia Econ. Finan. **17**, 113–120 (2014)

9. Deng, X., Li, J., Liu, E., Zhang, H.: Task allocation algorithm and optimization model on edge collaboration. J. Syst. Arch. **110**, 101778 (2020)

10. Dozier, C., Kondadadi, R., Light, M., Vachher, A., Veeramachaneni, S., Wudali, R.: Named entity recognition and resolution in legal text, pp. 27–43 (2010)

11. Eliot, L.: AI and legal argumentation: aligning the autonomous levels of AI legal reasoning (2020). https://doi.org/10.48550/ARXIV.2009.11180, arXiv:2009.11180

12. Elmabruk, R.: Judging the judges: examining supervisors assessment of unobservable skills in developed EFL teaching practice model. Eur. J. Educ. Stud. **7**, 157–182 (2020)

13. Flesca, S., Manco, G., Masciari, E., Pontieri, L., Pugliese, A.: Exploiting structural similarity for effective web information extraction. Data Knowl. Eng. **60**(1), 222–234 (2007). https://doi.org/10.1016/j.datak.2006.01.001, https://www.sciencedirect.com/science/article/pii/S0169023X0600022X, intelligent Data Mining

14. Kourlis, R.L., Singer, J.M.: Performance evaluation program for the federal judiciary. Denv. UL Rev. **86**, 7 (2008)

15. Kumar, G., Bhatia, P.K.: Impact of agile methodology on software development process. Int. J. Comput. Technol. Electron. Eng. (IJCTEE) **2**(4), 46–50 (2012)

16. Lupo, G., Bailey, J.: Designing and implementing e-justice systems: some lessons learned from EU and Canadian examples. Laws **3**(2), 353–387 (2014)

17. Manco, G., Masciari, E., Tagarelli, A.: A framework for adaptive mail classification. In: 14th IEEE International Conference on Tools with Artificial Intelligence, 2002, (ICTAI 2002), Proceedings, pp. 387–392 (2002). https://doi.org/10.1109/TAI.2002.1180829

18. Mansouri, A., Affendey, L.S., Mamat, A.: Named entity recognition approaches. Int. J. Comput. Sci. Netw. Secur. **8**(2), 339–344 (2008)

19. Masciari, E.: Trajectory clustering via effective partitioning. In: Andreasen, T., Yager, R.R., Bulskov, H., Christiansen, H., Larsen, H.L. (eds.) FQAS 2009. LNCS (LNAI), vol. 5822, pp. 358–370. Springer, Heidelberg (2009). https://doi.org/10.1007/978-3-642-04957-6_31

20. Minniti, L.: L'organizzazione del lavoro negli uffici giudiziari. In: L'organizzazione del lavoro negli uffici giudiziari, pp. 1000–1016 (2008)

21. Rule, C.: Online dispute resolution and the future of justice. Ann. Rev. Law Social Sci. **16**, 277–292 (2020)
22. Samanta, R., Ghosh, S.K., Das, S.K.: Swill-tac: skill-oriented dynamic task allocation with willingness for complex job in crowdsourcing. In: 2021 IEEE Global Communications Conference (GLOBECOM), pp. 1–6. IEEE (2021)
23. Sanh, V., Debut, L., Chaumond, J., Wolf, T.: Distilbert, a distilled version of bert: smaller, faster, cheaper and lighter (2019). https://doi.org/10.48550/ARXIV.1910.01108, arXiv:1910.01108
24. Skaff, M.K.: Implementing the electronic document management system at the local law enforcement by Iowa's judicial system: an application of the unified theory of acceptance and usage of technology. University of South Dakota (2016)
25. Wirtz, B.W., Weyerer, J.C., Geyer, C.: Artificial intelligence and the public sector-applications and challenges. Int. J. Public Adm. **42**(7), 596–615 (2019)
26. Zheng, Y., Yu, H., Cui, L., Miao, C., Leung, C., Yang, Q.: Smarths: an AI platform for improving government service provision. In: Proceedings of the AAAI Conference on Artificial Intelligence, vol. 32 (2018)

Ontology-Driven Parliamentary Analytics: Analysing Political Debates on COVID-19 Impact in Canada

Sabrina Azzi[1]([mail]) [ORCID] and Stéphane Gagnon[2] [ORCID]

[1] University of West of Scotland, Scotland, UK
`sabrina.azzi@uws.ac.uk`
[2] Université du Québec en Outaouais, Gatineau, Canada
`stephane.gagnon@uqo.ca`

Abstract. Parliamentary debates are usually published in Parliament's websites to allows citizens to be informed on the latest national debates, proposals and decisions. To enhance citizen experience and engagement, functionalities such as debates annotation and question answering are necessary. Annotating text requires semantic content and ontologies are known for their ability to describe a common vocabulary for a domain and can be a solid base for annotation and question answering. We report on an ongoing study to enhance parliamentary analytics using an ontology and knowledge graph to sharpen annotations and facilitate their query by end-users. As a salient case, a sample of debates are collected on the COVID-19 impact in Canada, as its complexity shows the relevance of using advanced knowledge representation techniques. We focused on the development of a new "Impact of COVID-19 in Canada Ontology" (ICCO) that provides contextualized semantic information on impact in numerous policy areas, as this ontology is entirely built from Canadian parliamentary debates. It has been evaluated and validated by experts. Our conclusion underscores the importance of integrating ontology-driven parliamentary analytics within the broader context of digital transformation in legislative institutions, and the need for new platforms supporting free and open Digital Humanities.

Keywords: Ontology · COVID-19 · Parliamentary debate

1 Introduction

In the latest three years, COVID-19 accelerated the phenomenon of digital transformation and become a pure necessity while it was before a just technological opportunity. Parliaments and governments known about their resistance and fear in adopting new technologies [1] have showed a great interest in incorporating innovative technologies in their processes and services towards electronic automation [2–5]. Moreover, with opening their non-confidential data online, enhancing citizens experience and engaging them is necessary to improve the quality of public policy-making and public services provisioning [6]. Websites are one of the means to promote parliamentary public engagement and being increasingly used by citizens makes it incumbent on parliaments to implement modern technologies for remaining relevant and friendly [7].

A. Kö et al. (Eds.): EGOVIS 2023, LNCS 14149, pp. 89–102, 2023.
https://doi.org/10.1007/978-3-031-39841-4_7

House of Commons of Canada (HoC) [8] recently invested in developing new technology functionalities in the context of legislative intelligent (LegisIntel) solutions project [3]. LegisIntel refers to Artificial Intelligence (AI) and analytical tools adopted by parliaments to enhance citizen experience in monitoring interrelations among diverse contents of parliamentary proceedings [3]. Text annotation that consists of reading and assigning labels to parts of the text is one of the solutions being studied within this project.

The present work is part of this project and consists of the first phase focused on developing an ontology to, among other, annotate text automatically, i.e., parliamentary debates. In fact, Canadian parliamentary debates are manually annotated, and annotation experts work daily on reviewing each new debate and annotate. This process is expensive, time consuming and usually prone to errors resulting from several factors such as user familiarity, personal motivation, etc. Annotating text requires semantic content and ontologies are known for their ability to describe a common vocabulary for a domain and define the meaning of the concepts and the relations between them. An ontology is defined as 'a formal and explicit specification of a shared conceptualization' [9]. Ontologies are also used for sharing, reusing, and integrating knowledge. Ontology development has been very active during the last five years in many sectors including electronic government and parliament due to the increasingly need of semantic, homogeneity, reasoning, etc.

COVID-19 is one of the effervescent topics in parliamentary discussions in the latest three years as it exponential spread has not only impacted public health [10] but also economy [11], labour market [12], education [13], etc. Using 'COVID-19 impact' as a keyword for search, search engine returned 955 long interventions published between 23 November 2021 and 31 March 2023 (see Fig. 1).

After discussions with information management team within the HoC, an ontology that covers the impact of COVID-19 in Canada is the best domain to start with to relieve annotation experts.

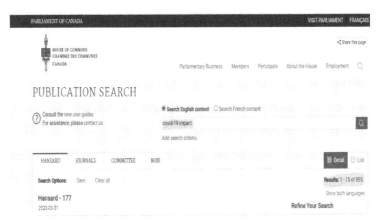

Fig. 1. Number of interventions obtained by the House of Commons of Canada website when searching the impact of COVID-19 in Canada.

This paper describes how we created the 'Impact of COVID19 in Canada Ontology' (ICCO). The next section outlines a related work. Materials and methods are presented in Sect. 2. Our ICCO development started with defining the domain and scope of ontology (phase 1). We used Canadian parliamentary debates as a primary knowledge source (phase 2) to build preliminary ICCO (phase 3). Parliamentary debate are formal discussions of proposals that the members of Parliament (MP) take in turns to speak. These debates are then transcribed, edited, corrected, and build up a Hansard. The preliminary ICCO has been subsequently refined through the reuse of the elements of other ontologies (phase 4). Finally, we evaluated ICCO using parliamentary debates and by involving domain experts (phase 5). Results and discussion are presented in Sect. 4 and conclusion is Sect. 5.

2 Related Work

Ontology-driven annotation of parliamentary debates depends heavily on reliable corpora and language features that can be accurately linked to ontology concepts and instances. For this reason, while ontologies have not yet been an extensive focus in this area, there has been important efforts by several Natural Language Processing (NLP) and computational linguistics researchers that pave the way for the present research. They can be classified in three groups depending on what language features they focus on: corpus integration, linguistic annotations, and argumentation mining.

At the corpus level, European teams have been among the most active in developing parliamentary corpora, especially with multilingual features and models [14]. A major example is ParlaMint, a multilingual annotated corpus available through the CLARIN.SI repository [15]. National parliamentary traditions can influence corpus contents and organization, and as such more targeted studies can allow more diverse feature extractions for AI applications. Many countries, again in Europe, have developed such corpora that enable advanced features, including Czech Republic [16], Denmark [17], Finland [18], Germany [19], Norway [20], Portugal [21], and Slovenia [22]. These corpus integration efforts include national language disambiguation and evolution analysis, while ensuring policy debates are contextualized in time and context.

At the linguistic level, annotations of parliamentary debates have offered the opportunity to use of linguistic metadata, which can be extracted reliably using NLP and deep learning methods, linking entities to knowledge graphs [23]. Typical annotations include speaker identification, debate context, keywords, and especially relations to various entities such as laws and policies. Unification of these features within a single corpus can help greatly reduce the semantic annotation task in later phases and provide a partial gold standard for classification tasks given the highly reliable curation process of these metadata. While AI applications are still limited in parliaments, there some countries have focused on semantic relation formalization and relationships extraction, while ensuring their reliable reuse, such as efforts in Chile [24–26], Greece [27–29], United Kingdom [30–32], and United States [33–35]. These studies rely on semantic technologies of varying maturity given they span a long period and have taken distinct approaches. But they share the common goal of simplifying annotation schemas, ensuring entities are accurately related to one another, facilitate government interoperability and legal-policy transparency, monitor legislative productivity and voting patterns, and used schemas to enrich debate analytics and search features.

At the discourse level, interactions between parliamentary speakers can be modelled through argument mining [36]. This approach involves developing typical patterns of speaker interactions, and identifying frequent associations among various linguistic features, whether words, entities, noun phrases, relationships, identities, and policy-related contexts. Taken within a flow of interactions, they can reveal how arguments take form, and how they influence legislative outcomes. Notable studies, relaying on varying degrees on computational linguistics, have been carried out in parliaments of Estonia [37, 38], the European Union [27, 39, 40], India [41], Ireland [42], Latvia [43], and Sweden [44]. Models of parliamentary arguments are increasingly recognizable, with potential to enrich ontologies with discourse patterns.

The present study builds upon recent advances at all three levels of parliamentary analytics. First, corpus development for our study relies on a rigorously, professionally curated bilingual and translated (English and French) corpus from the Canadian Hansard. We are therefore given assurances that the corpus and its keywords are a true gold standard with which our ontology annotations can be compared. Second, linguistic features of the debate samples rely again on existing relationships embedded in debate metadata. This allows us to draw COVID19-related debate contexts and ensure that its impacts are clearly identified as per the policy context of Canada. Third, as we plan pursuing discourse analysis in later stages, our debates will be annotated further using more complex semantic relationships within our new ontology, such as linking policy actions to public health outcomes, and therefore representing predefined discourse patterns as per the latest research on argument mining. As well, given a successful next stage in our research, we will focus our attention on measuring the use of "evidence-based policy making", i.e., the extent to which parliamentarians rely on complex fact-checking and knowledge understanding to ensure reliable decisions to fight the pandemic.

3 Materials and Methods

Many methodologies for building ontologies were proposed [45–49]. In [50], a new methodology was proposed and covers all the necessary phases to build an ontology of quality with providing guidance for the knowledge engineer. Unlike other methods that do not fully adhere to the Institute of Electrical and Electronics Engineers (IEEE) standards [51], this methodology supports a rigorous process of structuring the ontological concepts and facilitating modularization. This methodology has been previously used in building an ontology of quality that was evaluated by experts and also published [52].

The process of ICCO development is composed of five phases as illustrated in Fig. 2. During the process, we followed well-established Open Biological and Biomedical Ontology (OBO) Foundry principles [53] and the ARCHitecture for ONTological Elaborating (ARCHONTE) method to create preliminary ICCO.

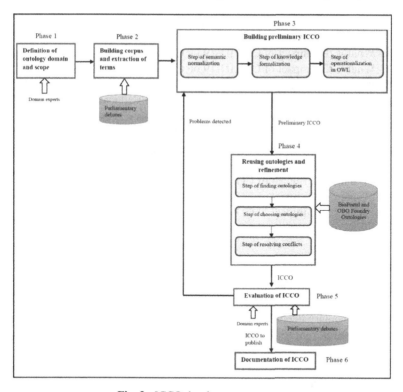

Fig. 2. ICCO development process.

3.1 Phase 1: Definition of Ontology Domain and Scope

This phase consists of establishing the domain of interest of the ontology to model and its scope. It is done by consulting domain experts within the HoC. First, two fundamental questions must be answered:

1. What is the domain that the ontology will cover?
2. What is main purpose of developing the ontology?

Second, functional requirements for the ontology or questions that an ontology should be able to answer are defined. In order to support this, competency questions (CQs) established by domain experts can be used. Indeed, these CQs are presented in the form of informal queries that a particular community of users believes that the ontology should answer.

We defined the domain of ontology and scope after discussions with information management experts within the HoC. The impact of COVID-19 in Canada, which is a very of effervescent topic in parliamentary debates in the latest three years, was identified. The following CQs were established by information management experts withing the HoC:

1) CQ1- What is COVID-19?
2) CQ2- What are the different sectors impacted by COVID-19 in Canada?

3) CQ3- What is the impact of COVID19 on Canada's economy?
4) CQ4- What is the impact of COVID-19 on Canadian mental health?
5) CQ5- What is the impact of COVID-19 on Canadian public health?
6) CQ6- What is the impact of COVID-19 on Canada's healthcare system?
7) CQ7- What is the impact of COVID-19 on Canada's financial system (banking…)?
8) CQ8- What is the impact of COVID-19 on Canada's education?
9) CQ9- What is the impact of COVID-19 on Canada's real estates?
10) CQ10- What is the impact of COVID-19 on Canada's labour market?
11) CQ11- What is the impact of COVID-19 on Canada's immigration?
12) CQ12- What is the impact of COVID-19 on Canada's tourism?
13) CQ13- What is the impact of COVID-19 on Canada's legislation?
14) CQ14- What is the impact of COVID-19 on Canada's media?
15) CQ15- What is the impact of COVID-19 on Canada's public safety?
16) CQ16- What is the impact of COVID-19 on Canada's federal-provincial relations?

3.2   Phase 2: Building Corpus and Extraction of Terms

In this phase, we extracted and identified relevant terms to the ontology domain defined in phase 1, from a corpus of knowledge. The corpus of knowledge is constituted of publicly available parliamentary debates in Hansard [54] latter is a verified and accurate record rather than verbatim transcript. Redundancies are removed, recognizable mistakes are corrected, and interjections omitted unless responded to by the principal speaker. Our corpus of knowledge is composed of 168 members of Parliament interventions from December 2021 to March 2022.

We used Text2Onto [55] tool that offers good results to extract terms from the debates. Text2Onto has been developed to support the acquisition of ontologies from textual documents. It offers three extraction algorithms calculating the following measures: Relative Term Frequency (RTF), Term Frequency Inverted Document Frequency (TFIDF), Entropy, and the C-value/NC-value. TFIDF gives the best results because it evaluates how relevant a term is to a document in a collection of documents.

3.3   Phase 3: Building Preliminary ICCO

This phase was dedicated to the building of a preliminary ICCO (p-ICCO) by following the ARCHONTE methodology. The strong point of ARCHONTE is the recommendation of semantic commitment in the design process. In other words, explaining the meaning of each of the ontology concepts, the similarities and differences that the concept has with those close to it using natural language. The ontology structure is similar to a tree structure, which facilitates the determination of the meaning that a concept has according to its relative position. ARCHONTE is composed of three steps: 1) semantic normalization; 2) knowledge formalization; and 3) toward a computational ontology. They are elaborated below.

Step of Semantic Normalization

The objective of this step is to reach a semantic agreement about the meaning of the concept labels. We applied differential principles on the set of candidate concepts chosen previously. According to the differential paradigm proposed by Bachimont [56], the meaning of a concept (or a node, since an ontology is structured as a tree) is determined by its closest neighbors: parent and its siblings, *i.e.*, its position in the ontological hierarchy based on terminological structure. Similarities and differences of each concept concerning its parent and its siblings are expressed in natural language. Four differential principles are distinguished: 1) similarity with each parent; 2) similarity with each sibling; 3) difference with each sibling; 4) difference with each parent. The result of this step is a so-called differential ontology. We applied these semantic principles to create a hierarchy of concepts and relationships organized according to their similarities and differences. For example: *primary school, secondary school* and *university* are sibling concepts because of their similarity (education), and at the same time, they are different concepts because represent different level of education.

Step of Knowledge Formalization

The objective of this step is to link each concept to a set of objects that allows defining new concepts and relations through set operations. During this step, we specified the arity and domains of the relations, in correspondence with the intended models. We added some logical axioms to constrain the domains of the relations. The result of this step was an ontology that covered differentials concepts, relations, instances, and axioms, such as subclass ones, for the impact of COVID-19 in Canada domain.

Step of Operationalization

The objective of this step is implementing the ontology in an operational language of knowledge representation such as OWL. We used Protégé [57], we operationalized p-ICCO in OWL language because it perfectly meets our needs in terms of expressiveness and manageability. At this last step of the ARCHONTE method, we obtained the final version of p-ICCO.

3.4   Phase 4: Reusing Ontologies and Refinement

In this phase, we added and detailed concepts in p-ICCO. For this purpose, we looked for other relevant ontologies. These ontologies had to be evaluated for the suitability of their content coverage and the depth of knowledge they represent. Unfortunately, no ontology responded to the requirements of ICCO except DOID [58] that covers human diseases. We used Wikipedia to enrich concepts with definitions and synonyms.

3.5   Phase 5: Evaluation of ICCO

Ontology evaluation is the process of assessing the quality of an ontology concerning a set of evaluation criteria and its adequacy for being used in a specific context for a specific goal. A classification of ontology evaluation was proposed by Brank and his colleagues in [59], and it consists of four categories: 1) approaches that use the target ontology in an application and evaluate the application results [60]; 2) approaches that compare the target ontology to a gold standard [61]; 3) those that use data sources about a specific domain [62]; and 4) approaches that recommend a manual assessment by domain experts according to a set of criteria [63].

Internal Consistency and Adherence to Standard Ontology Practices
We used Pellet reasoner integrated with PROTÉGÉ to check the internal consistency of ICCO, and no errors were detected. We manually verified the full adherence of ICCO to OBO Foundry guidelines and Cimino's desiderata [64].

Evaluation Using Dependent Domain Sources
We used new 126 member of Parliament interventions from April 2022 to June 2022. The goal of this step was to allow us to verify each concept from an intervention that may refer to the domain, whether it is addressed in the ontology or not. The concepts concerned by the annotation process include adjectives, adverbs, proper nouns, nouns, verbs, verbal phrases, nominal phrases, adjectival phrases, and participatory phrases. At the end of this step, 39 new concepts were added to ICCO. This step was manual.

We also evaluated ICCO by using an annotator developed within our team. The annotator was connected to ICCO and for each intervention, it highlights concepts related to the impact of COVID-19 in Canada by providing definitions. All needed concepts were annotated.

Evaluation by Domain Experts
The evaluation is done by two domain experts, from information and management team within the HoC, to assess how well the ontology meets a set of predefined requirements. They recommended to add more relations between financial system and economy and axioms. For example: *deficit*, *government spending*, and *inflation* that had *financial system* as parent should also have *economy* as parent.

3.6 Phase 6: Documentation of ICCO

We annotated every concept in ICCO and used OWLDoc to document the ontology. Every phase of the building process is described in this work.

4 Results and Discussion

ICCO contains 501 classes, 41 object properties, 4069 axioms, 56 individuals (instances) and 19 annotation properties. Figure 3 shows an overview of ICCO. Figure 4 illustrates the main classes of ICCO demonstrating the hierarchy of class tourism. The object properties are presented in Fig. 5. For example, the object property *provide* asserts the relation between a vaccine and an immunity *vaccine **provide** immunity.*

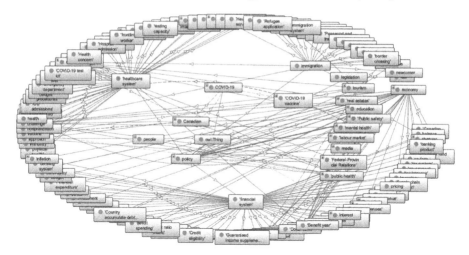

Fig. 3. ICCO overview.

The resulted ICCO covers the impact of COVID on education, healthcare system, public health, public safety, mental health, real estates, labor market, tourism, media, legislation, federal-provincial relations, immigration, economy, and financial system.

The phase 3 was the most consuming during the process of ICCO development. This phase required semantic commitment in analyzing each concept regarding its siblings and parents. The phase 4 was the quickest one as no ontology responded to ICCO requirements. Therefore, we did not benefit from ontology reuse which would have enrich ICCO with concepts and definitions, synonyms, etc., and increase reusability and interoperability with existing ontologies.

For the interoperability and reusability purposes, all ICCO concepts were annotated with standard identifiers formatted as recommended by OBO Foundry, synonyms, and definitions collected as well from Wikipedia, where such information is available.

The main limitation of ICCO is the lack of modular structure. A modular ontology design offers numerous benefits, such as easy reuse of a single component in other ontologies, good reasoning, facilitating ontology understanding, reduced complexity, to name a few. While there were attempts to define how to select "the best modular ontology" [65], currently there are no tools supporting the creation of modular ontologies. This limitation applies to the majority of ontologies publicly available.

Fig. 4. ICCO main classes and tourism hierarchy   **Fig. 5.** ICCO object properties

5   Conclusion

This study has presented a new ontology specifically targeted for integration within an ontology-driven parliamentary debate annotation platform. The new "Impact of COVID-19 in Canada Ontology" (ICCO) provides contextualized semantic information on impact in numerous policy areas, as this ontology is entirely built from Canadian parliamentary debates. We began with a review of the latest advances in the field of legislative debate analytics, classifying studies at the levels of corpus integration, linguistic annotation, and argument mining. We followed with an outline of the materials and methods of our study, namely the project lifecycle for ontology design, implementation, and evaluation. Findings from this effort were then discussed within the context of parliamentary and legislative intelligence.

The potential impact of ontology-driven intelligence can be positioned within the broader context of digital transformation in parliaments [66], where AI offers promising opportunities, enabling seamless integration across parliamentary and government processes. Ontologies and knowledge graphs are essential assets in formalizing artificial intelligence capabilities in various operational contexts. Given that parliaments are primarily discursive environments, they present less constraints in data, information, and knowledge modelling. These solutions can be improved by integrating specialized ontologies to provide contextualized policy-specific analytics, such as the one presented here on COVID19 impact in Canada. Their systematic reuse, and continuous enrichment, can also serve to rapidly enhance user experience in legislative intelligence applications.

References

1. Meijer, A.: E-governance innovation: Barriers and strategies. Gov. Inf. Q. **32**(2), 198–206 (2015)
2. Leventis, S., Anastasiou, V., Fitsilis, F.: Application of enterprise integration patterns for the digital transformation of parliamentary control. In: Proceedings of the 13th International Conference on Theory and Practice of Electronic Governance (2020)
3. Gagnon, S., Azzi, S.: Semantic Annotation of Parliamentary Debates and Legislative Intelligence Enhancing Citizen Experience. Springer International Publishing, Cham (2022). https://doi.org/10.1007/978-3-031-12673-4_5
4. Al-Mushayt, O.S.: Automating E-government services with artificial intelligence. IEEE Access **7**, 146821–146829 (2019)
5. Engstrom, D.F., et al.: Government by algorithm: Artificial intelligence in federal administrative agencies. NYU School of Law, Public Law Research Paper, pp. 20–54 (2020)
6. Purwanto, A., Zuiderwijk, A., Janssen, M.: Citizen engagement with open government data: a systematic literature review of drivers and inhibitors. Int. J. Electronic Governm. Res. (IJEGR) **16**(3), 1–25 (2020)
7. Serra-Silva, S.: How parliaments engage with citizens? Online public engagement: a comparative analysis of Parliamentary websites. J. Legislative Stud. **28**(4), 489–512 (2022)
8. House of Commons of Canada (2023). https://www.ourcommons.ca/en (Accessed 12 April 2023]
9. Studer, R., Benjamins, V.R., Fensel, D.: Knowledge engineering: Principles and methods. Data Knowl. Eng. **25**(1–2), 161–197 (1998)
10. Levitt, E.E., et al.: Public health guideline compliance and perceived government effectiveness during the COVID-19 pandemic in Canada: Findings from a longitudinal cohort study. The Lancet Regional Health-Americas **9**, 100185 (2022)
11. Beland, L.P., et al.: The short-term economic consequences of COVID-19: Occupation tasks and mental health in Canada. Canad. J. Econom./Revue canadienne d'économique **55**, 214–247 (2022)
12. Lemieux, T., et al.: Initial impacts of the COVID-19 pandemic on the Canadian labour market. Can. Public Policy **46**(S1), S55–S65 (2020)
13. Buckner, E., Zhang, Y., Blanco, G.L.: The impact of COVID-19 on international student enrolments in North America: Comparing Canada and the United States. High. Educ. Q. **76**(2), 328–342 (2022)
14. Gartner, R.: Using structured text corpora in parliamentary metadata language for the analysis of legislative proceedings. Digital Humanities Q. **12**(2) (2018)
15. Erjavec, T., et al.: The ParlaMint corpora of parliamentary proceedings. Lang. Resou. Evaluat. (2022)
16. Kratochvíl, J., Polák., P., Bojar, O.: Large corpus of czech parliament plenary hearings. In: 12th International Conference on Language Resources and Evaluation, LREC 2020. European Language Resources Association (ELRA) (2020)
17. Kirkedal, A., Stepanovic, M., Plank, B.: FT SPEECH: Danish parliament speech corpus. In: 21st Annual Conference of the International Speech Communication Association, INTERSPEECH 2020, International Speech Communication Association (2020)
18. Mansikkaniemi, A., Smit, P., Kurimo, M.: Automatic construction of the Finnish parliament speech corpus. In: 18th Annual Conference of the International Speech Communication Association, INTERSPEECH 2017, International Speech Communication Association (2017)
19. Blätte, A., Blessing, A.: The Germaparl corpus of parliamentary protocols. In: 11th International Conference on Language Resources and Evaluation, LREC 2018, European Language Resources Association (ELRA) (2019)

20. Lapponi, E., Søyland, M.G., Velldal, E., Oepen, S.: The talk of Norway: a richly annotated corpus of the Norwegian parliament, 1998–2016. Lang. Resour. Eval. **52**(3), 873–893 (2018). https://doi.org/10.1007/s10579-018-9411-5

21. Almeida, P., Marques-Pita, M., Gonçalves-Sá, J.: PTPARL-D: An annotated corpus of forty-four years of Portuguese parliamentary debates. Corpora **16**(3), 337–348 (2021)

22. Fišer, D., Ljubešić, N., Erjavec, T.: Parlameter - A corpus of contemporary slovene parliamentary proceedings. Prispevki za Novejso Zgodovino **59**(1), 70–98 (2019)

23. Lu, F., Cong, P., Huang, X.: Utilizing textual information in knowledge graph embedding: a survey of methods and applications. IEEE Access **8**, 92072–92088 (2020)

24. Cifuentes-Silva, F., et al.: Describing the nature of legislation through roll call voting in the chilean national congress, a linked dataset description. Semantic web (2021)

25. Cifuentes-Silva, F., Fernández-Álvarez, D., Labra-Gayo, J.E.: National budget as linked open data: new tools for supporting the sustainability of public finances. Sustainability **12**(11), 4551 (2020)

26. Cifuentes-Silva, F., Labra Gayo, J.E.: Legislative document content extraction based on semantic web technologies. In: Hitzler, P., et al. (eds.) ESWC 2019. LNCS, vol. 11503, pp. 558–573. Springer, Cham (2019). https://doi.org/10.1007/978-3-030-21348-0_36

27. Loukis, E., et al.: Argumentation systems and ontologies for enhancing public participation in the legislation process. In EGOV (2007)

28. Loukis, E. and A.L. Xenakis. A methodology for ontology-based knowledge-level interoperability among parliaments. In: 15th Americas Conference on Information Systems 2009, AMCIS (2009)

29. Figgou, L., Andreouli, E.: Appeals to consensus and partisan politics in parliamentary discourse on the pandemic. Political Psychol. (2022)

30. Nanni, F., et al.: Semantifying the UK Hansard (1918--2018). IEEE Press (2019)

31. Abercrombie, G., Batista-Navarro, R.: Semantic Change in the Language of UK Parliamentary Debates. Association for Computational Linguistics (2019)

32. Coole, M., Rayson, P., Mariani, J.: Unfinished Business: Construction and Maintenance of a Semantically Tagged Historical Parliamentary Corpus, UK Hansard from 1803 to the present day. European Language Resources Association (2020)

33. Ballard, A.O.: Bill text and agenda control in the US congress. J. Polit. **84**(1), 335–350 (2022)

34. Atkinson, M.L., Mousavi, R., Windett, J.H.:Detecting diverse perspectives: using text analytics to reveal sex differences in congressional debate about defense. Political Res. Q. (2022)

35. Xing, Z., Hillygus, S., Carin, L.: Evaluating U.S. Electoral representation with a joint statistical model of congressional roll-calls, legislative text, and voter registration data. In: 23rd ACM SIGKDD International Conference on Knowledge Discovery and Data Mining, KDD 2017. Association for Computing Machinery (2017)

36. Lawrence, J., Reed, C.: Argument mining: A survey. Comput. Linguist. **45**(4), 765–818 (2019)

37. Koit, M.: How are the members of a parliament arguing? analysis of an argument corpus. in 13th International Conference on Agents and Artificial Intelligence, ICAART 2021. SciTePress (2021)

38. Koit, M.: Annotating arguments in a parliamentary corpus: An experience. In: 12th International Conference on Knowledge Discovery and Information Retrieval, KDIR 2020 - Part of the 12th International Joint Conference on Knowledge Discovery, Knowledge Engineering and Knowledge Management, IC3K 2020. SciTePress (2020)

39. Defrancq, B.: The european parliament as a discourse community: its role in comparable analyses of data drawn from parallel interpreting corpora. Interpreters Newsletter **23**, 115–132 (2018)

40. Calzada-Pérez, M.: Researching the european parliament with corpus-assisted discourse studies from the micro- and macro-levels of text to the macro-context. Revista Espanola de Linguistica Aplicada **30**(2), 465–490 (2018)

41. Sen, A., et al.: Studying the discourse on economic policies in India using mass media, social media, and the parliamentary question hour data. In: 2019 ACM SIGCAS Conference on Computing and Sustainable Societies, COMPASS 2019. Association for Computing Machinery, Inc. (2019)

42. Burroughs, E.: The discourse of controlling "illegal immigration" in irish parliamentary texts. J. Lang. Politics **14**(4), 479–500 (2015)

43. Skulte, I., Kozlovs, N.: Discourse on safety/security in the parliamentary corpus of latvian saeima. In: 5th Conference Digital Humanities in the Nordic Countries, DHN 2020. CEUR-WS (2020)

44. Eide, S.R.: The swedish poligraph: A semantic graph for argument mining of swedish parliamentary data. In: 6th Workshop on Argument Mining, ArgMining 2019, collocated with ACL 2019. Association for Computational Linguistics (ACL) (2019)

45. Gruninger, M. and M.S. Fox. The design and evaluation of ontologies for enterprise engineering. In: Workshop on Implemented Ontologies, European Conference on Artificial Intelligence (ECAI) (1994)

46. Uschold, M., King, M.: Towards a methodology for building ontologies. Citeseer (1995)

47. Noy, N.F., McGuinness, D.L.: Ontology development 101: A guide to creating your first ontology. 2001, Stanford knowledge systems laboratory technical report KSL-01–05 and (2001)

48. Staab, S., et al.: Knowledge processes and ontologies. IEEE Intell. Syst. **16**(1), 26–34 (2001)

49. Bachimont, B., Isaac, A., Troncy, R.: semantic commitment for designing ontologies: a proposal. In: Gómez-Pérez, A., Benjamins, V.R. (eds.) EKAW 2002. LNCS (LNAI), vol. 2473, pp. 114–121. Springer, Heidelberg (2002). https://doi.org/10.1007/3-540-45810-7_14

50. Azzi, S.: Nouvelle méthodologie de construction d'ontologies médicales: cas d'étude: diagnostic de la pneumonie. Université du Québec en Outaouais, p. 135 (2021)

51. Schultz, D.J., IEEE standard for developing software life cycle processes. IEEE Std, pp. 1074–1997 (1997)

52. Azzi, S., Michalowski, W., Iglewski, M.: Developing a pneumonia diagnosis ontology from multiple knowledge sources. Health Inform. J. **28**(2), 14604582221083850 (2022)

53. Open Biological and Biomedical Ontology Foundry (2023). https://obofoundry.org/princi ples/fp-000-summary.html (Accessed 3 April 2023]

54. Hansard (2023). https://www.ourcommons.ca/PublicationSearch/en/?PubType=37 (Accessed 9 April 2023)

55. Cimiano, P., Völker, J.: text2onto. In: Montoyo, A., Muñoz, R., Métais, E. (eds.) Natural Language Processing and Information Systems, pp. 227–238. Springer, Berlin (2005). https://doi.org/10.1007/11428817_21

56. Bachimont, B., Isaac, A., Troncy, R.: Semantic commitment for designing ontologies: a proposal. In: International Conference on Knowledge Engineering and Knowledge Management. Springer (2002). https://doi.org/10.1007/3-540-45810-7_14

57. PROTÉGÉ (2023) (Accessed 20 April 2023)

58. Human Disease Ontology (2023). https://bioportal.bioontology.org/ontologies/DOID (Accessed 2023)

59. Brank, J., Grobelnik, M., Mladenic, D.: A survey of ontology evaluation techniques. In: Proceedings of the conference on data mining and data warehouses (SiKDD 2005). Citeseer, Ljubljana, Slovenia (2005)

60. Porzel, R., Malaka, R.: A task-based approach for ontology evaluation. In: ECAI Workshop on Ontology Learning and Population, Valencia, Spain. Citeseer (2004)

61. Maedche, A., Staab, S.: Measuring similarity between ontologies. In: International Conference on Knowledge Engineering and Knowledge Management. Springer (2002). https://doi.org/10.1007/3-540-45810-7_24

62. Brewster, C., et al.: Data driven ontology evaluation (2004)

63. Lozano-Tello, A., Gómez-Pérez, A.: Ontometric: A method to choose the appropriate ontology. J. Database Managem. (JDM) **15**(2), 1–18 (2004)

64. Cimino, J.J.: Desiderata for controlled medical vocabularies in the twenty-first century. Methods Inf. Med. **37**(4–5), 394–403 (1998)

65. Kumar, S., Baliyan, N.: Quality evaluation of ontologies. In: Semantic Web-Based Systems. SCS, pp. 19–50. Springer, Singapore (2018). https://doi.org/10.1007/978-981-10-7700-5_2

66. Koryzis, D., et al.: Parltech: Transformation framework for the digital parliament. Big Data Cognitive Comput. **5**(1) (2021)

Hybrid AI Analysis of the Drug Micro-trafficking in Italy

Salvatore Sapienza(✉) 📵

Department of Legal Studies, University of Bologna, Bologna, Italy
salvatore.sapienza@unibo.it

Abstract. This paper analyses factual and legal aspects on the *quantum* of criminal sanctions in cases of drug micro-trafficking. This case study considers the Italian jurisdiction, which contemplates two legal qualifications of drug dealing, i.e., "minor" and "non-minor" offences. As a case-by-case analysis is required to courts of merits in deciding how the offence has to be legally qualified, the study aims to identify trends and to cast light and provide explanations on the judicial decision-making about the legal qualification of a set of facts. The study emphasizes the importance of combining criminal judgment with innovative tools to facilitate the work of judges and allow information-based studies about jurisprudential trends.

Keywords: Legal Data Analysis · Digital Justice · Micro-trafficking · Digital transformation · Data Science in justice

1 Introduction

When a criminal judge of merits is tasked with evaluating specific facts and applying syllogistic reasoning to relevant criminal facts, transparency, uniformity, and coherence must be ensured in the decision-making process. Although Italy is a Civil Law country, the principles of predictability of criminal sanctions and formal equality require that some degree of relevance is given to past cases. The similarity between the factual elements of the current case and previous ones, as well as the legal proximity of relevant institutions, can determine the level of closeness between analogous cases. To uphold equality in criminal justice, it would be necessary to provide judges with information that can quickly help them to navigate legal and factual elements of a case, enabling them to make well-informed decisions about the quantum of sanctions. Moreover, such information could be helpful to identify latent trends or emerging patterns in jurisprudence, thus being helpful for legal scholars, practitioners, students, and the whole judicial system.

The urgency of the issue arises particularly in cases of micro-trafficking that are often brought to the judge's attention. Micro-trafficking involves easily identifiable typifying elements such as criminal conduct, the object of the crime, the criminally relevant quantity of traded substance, and the subjective element.

Taken together, these factors make micro-trafficking a suitable scenario for the automated analysis of legal texts. The repetitive nature of these cases provides an ideal testing ground for combining criminal judgment with innovative tools aimed at the analysis of rulings.

The frequency of micro-trafficking cases requires the semi-automated identification of constituent elements of the criminological type directly from previous case law texts. Once identified, these elements can be further extracted, refined, and analysed using legal data analytics tools to facilitate the work of judges facing similar cases. This pipeline constitutes the object of this introductory study. By analysing a corpus of rulings, this study identifies trends and provide explanations on judicial decision-making. Moreover, the goals of this study do not entail "predictive justice" or "judicial profiling". Although some of the models presented hereafter have the ability to predict the outcome of the case, we opted to refrain from predicting the judgements as this would have led to ethically-questionable results.

The paper is structured as follows. Section 2 identifies some related works which constitute the background of this study. Section 3 discusses the research question and the methodology adopted in this paper. Section 4 presents the legal background of drug micro-trafficking in the Italian legislation and case law. Section 5 discusses how the dataset has been built, whereas Sect. 6 presents some findings. A discussion (Sect. 7) on such findings anticipates some final remarks (Sect. 8).

2 Related Works

In his presentation at the seventeenth International Conference on Artificial Intelligence and Law (ICAIL'19), Verheij [1] argued that AI systems in the legal domain should be seen as "hybrid critical discussion systems" (emphasis added). According to this approach, AI systems construct and evaluate hypothetical perspectives to find satisfactory solutions, with the aim of assisting legal professionals in making informed decisions and providing accurate legal advice, not necessarily relying solely on AI systems. Legal analytics through Artificial Intelligence consists of several approaches [2]. The field of computational law aims to address complex problems, such as legal interpretation [3], argument mining [4–6], rule extraction [7], and the management of temporal aspects of legal documents [8,9]. To achieve these goals, researchers have developed two main categories of computational approaches: legal expert systems and legal text analysis [2,10,11].

Automated legal expert systems use rule-based systems, case-based reasoning, and machine learning algorithms to provide legal advice or decision-making support. Rule-based systems mimic logical reasoning used by legal experts for interpretation and rule extraction. Legal text analysis uses natural language processing and machine learning to extract information from legal documents and provide insights into legal cases. Techniques such as argument mining and classification are used to identify arguments and categorize legal texts. The overviews provided in [2, p. 73, 260] clarify the scope of legal information retrieval from statutory text (e.g., prohibitions, obligations, and so forth and from rulings (e.g., arguments). The scenario presented in this study differs from this literature because it aims at extracting factual information from rulings. Therefore, some adaptations to the methods presented in the related works are necessary. Such changes are discussed in the section below.

3   Research Question and Methodology

The research methodology used in this study is a hybrid approach that combines unsupervised and supervised learning experiments on a legal corpus. This methodology involves several steps, which are illustrated in Fig. 1. The first step is a legal analysis that identifies an interesting or controversial research question for legal scholars. In this study, the research question is: "What factual legal factors, both quantitative and qualitative, contribute to the classification of drug-dealing actions as "minor" or "non-minor" crimes?" As we will see in the next section, this question is significant for legal scholars seeking to provide answers to the demand for legal certainty in cases where the classification of the offence is uncertain. While the motivation behind this question is primarily legal, a legal-informatics approach can be useful in formulating or validating legal hypotheses. Then, following the definition of the question, legal research aims to identify a) legal "signals" in the language that are relevant to address the research question and b) what legally-qualified factors can be of interests. For instance, a) the occurrence of the string <number + unit of measure + "of" + illegal substance> is frequent, and b) it can be used to investigate whether correlations with legal factors such as the applicability of drug trafficking legislation.

Legal signals are formalised in a light taxonomy, which helps navigating the signals and their relationship. Such taxonomy does not consists of all the elements of legal ontologies (e.g., restrictions and limitations), yet it allows a broad understanding of how the elements that are necessary to perform the analysis are related. Thus, where applicable, MeLOn methodology for legal ontologies is used to design such light taxonomy [12]. Then, unsupervised experiments are carried out to understand the extent to which the dataset contains features that allow the automated retrieval of information. Extracted factors are analysed and, if necessary, the light taxonomy is refined to include the newly-extracted factors. Legally-relevant factors are extracted from the text. Such factors can be extracted manually (e.g., in [13]) by means of annotating legal documents, or automatically (e.g., in [14]). This study adopted both the approaches by means of regular expressions and manual extraction/validation of the legal factors. Finally, following the measurement of the performance by metrics, these results are interpreted in a legal sense. Limitations of the method are discussed and mitigating solutions are proposed for further refinements. An answer to the research question is provided after the twofold (technical and legal) validation.

The methodological pillars of the proposed method include exaplainability and knowability [15] of the results. This implies that attention is placed on

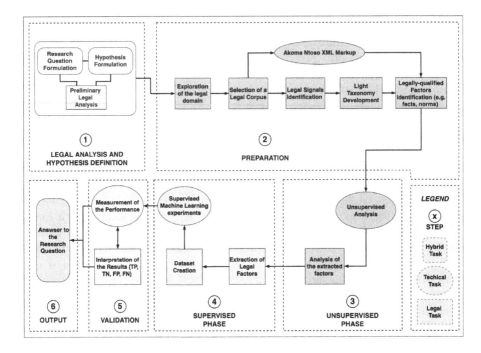

Fig. 1. Hybrid Research Methdology

allowing the correct interpretation of the results from a legal perspective. This is in line with recent studied in legal informatics, which have correctly pointed out the necessity of explaining conclusions in legal analysis software [16,17].

4   The Legal Background in the Italian Micro-trafficking

Article 73 of the Presidential Decree 309/199n outlines the penalties for drug sale and trafficking. If someone grows, makes, sells, transports, or trades any narcotic or psychotropic substance without the necessary authorization, they will be sentenced to imprisonment for a period of 6 to 20 years and a fine ranging from €26,000 to €260,000. In 2019, the Italian Constitutional Court has lowered the minimum sentence from 8 years to 6 years by considering the 8-year minimum disproportionate to the offence[1], also in the light of the offence range of more serious crimes like personal injuries or murder attempts [18]. Therefore, great attention is placed on the offensiveness of the drug-dealing action when the case is decided and the defendant is sentenced.

Alongside the provision against serious drug trafficking, a specific mitigating circumstance is the one provided in Article 73(5), which states that if the actions committed, the way they were carried out, or the substances involved are considered "minor", the punishment for violating the law will be imprisonment for

---

[1] Italian Constitutional Court Ruling 40 of 2019.

a period of 1 to 6 years and a fine ranging from €3,000 to €26,000. Given the differences in the legal regime, legal reasoning is necessary to correctly qualify the facts as a "minor" or "non-minor" offence[2]. The significance of "borderline cases", i.e., those "placed in a grey zone between the two types of offence"[3] is relevant considering the *nullum crimen sine lege* principle, according to which individuals should now the possible no conduct shall be held criminal unless it is specifically described in the behaviour circumstance element of a penal statute [19].

Crucially, the legislation does not specify any threshold separating "minor" and "non-minor" facts. The evaluation shall be carried out in courts of merits. The Italian Court of Cassation has stated that a) a case by case analysis has to be carried out by each judge and, b) the attribution of the "minor" or "non-minor" charge has to be grounded on quantitative and qualitative elements of the facts[4].

The risk is that "[t]he extent of the penalty gap inevitably conditions the overall assessment that the trial judge must make in order to ascertain the minor extent of the fact (deemed necessary by the Court of Cassation, joint criminal sections, ruling n. 51063/2018), with the risk of giving rise to punitive inequalities, in excess or in default, as well as unreasonable application discrepancies in a significant number of conducts"[5]. The problem lies both on the extent of the penalty range and on the twofold regime of "minor" and "non-minor", which requires a necessary qualitative and quantitative assessment. However, such evaluation is not straightforward as it seems. Qualitative and quantitative elements may be defined in an (almost) infinite number of possibilities, which include information related to the illegal substance as well as factual elements linked to *mens rea* and *actus rei* such as the habituality of the charged person.

Moreover, other elements may have an impact over the final decision ("minor" or "non-minor") despite not being directly mentioned by the law. For instance, a condition of recidivism or the procedure (e.g., shortened proceedings, pleas) might influence the decision. Therefore, it is necessary to make a selection of the features that might be relevant. To preserve the legal integrity of the study, only certain elements which are undoubtedly relevant to the illegal substance have been considered, such as the quantity/quality and purity indicators of the

---

[2] As one of the reviewers has correctly pointed out, this differentiation might lead to ambiguity in the definition of "non-minor" offences, which may encompass every legal qualification other than "minor". However, this lexical choice was done on purpose to mimic the Italian jurisdiction, which contemplates two different charges ("minor" and "non-minor" offences), and an aggravating circumstance for "serious" offences without a specific charge. Accordingly, the problem has been approached as a binary classification problem.

[3] Italian Constitutional Court Ruling 40 of 2019 §5.2.

[4] Court of Cassation, ruling n. 51063/2018 issued by the joint criminal sections, §5; Court of Cassation, 7th criminal section, ruling n. 6621/2019; Court of Cassation; 7th criminal section, ruling n. 3350/2019; Court of Cassation, 4th criminal section, ruling n. 2312/2019.

[5] Italian Constitutional Court Ruling 40 of 2019 §5.2.

substance. Alongside them, procedural elements are used to investigate potential correlations with the final legal qualification of the facts ("minor"/"non-minor").

In 2022, a study carried by the Italian Court of Cassation found out that the judges of its Sixth Section consider the "minor" qualification to be correctly applied by looking also (and not exclusively) to the quantitative elements related to the absolute weight of the traded substance[6] [20]. Crucially, judges from the Court of Cassation do not decide as courts of merits, but ultimately provide the correct interpretation of the law in case of jurisprudential debate or uncertainty on points of law. Its interpretation becomes highly influential, yet an in-depth analysis is required to judges on the merits in cases such as micro-trafficking. Moreover, the adaptation from the courts of merits to such "minor"/"non-minor" thresholds identified by the Court of Cassation might take some time. In the meanwhile, it is worth reconstructing previous decisions rather than forecasting future outcomes. In fact, the soundness of any forecast on previous cases would be seriously hindered by the new trend emerging in the rulings issued by the Court of Cassation.

5   Dataset Composition

The composition of the dataset is the following. From a corpus of rulings (N = 340), a subset (N = 88) of rulings in which the suspect has been found guilty has been extracted. In 49 rulings (.558), the offender has been charged with "minor" drug dealing, whereas in the remaining 39 (.442), the offence has been considered "non-minor". ll the rulings are anonymised prior to further processing in compliance with the applicable laws and codes of ethics. Table 1 describes the data available in the dataset, a short description of the information available, the data type, the unit of measure (when available) and how the information has been extracted.

The issue of managing legal references (legislation and precedents) is addressed by modeling legal knowledge through an international XML standard, Akoma Ntoso [21], adopted by OASIS and used by many institutions for marking up and sometimes even drafting their own legal documents[7] The benefits of using Akoma Ntoso lie in enriching the digitized text with precise legal references (articles, commas, statutes, etc.) and point-in-time information, allowing for a secure extraction of provisions and precedents. Unlike other models, this enrichment preserves the legal meaning of the original text and serves as useful metadata for automatic processing while respecting the semantics attributed by humans. Akoma Ntoso supports judicial documents by identifying `<introduction>`, `<background>` (facts), `<motivation>` and `<decision>` elements within the `<judgementBody>` of the cases alongside metadata [22].

---

[6] These thresholds are the following 1) 23.66 g for cocaine; 2) 28.4 g for heroin; 3) 108.3 g for marijuana; 4) 101.5 g for hashish.

[7] See   http://docs.oasis-open.org/legaldocml/   akn-core/v1.0/os/part1-vocabulary/ akn- core-v1.0-os-part1-vocabulary.html, last accessed on 19 April 2023.

Table 1. Information types extracted from the rulings, alongside a short description by, the data type, the unit of measure (when available) and the extraction method

Information	Description	Data Type	Unit of Measure	Extraction
substance	The illegal substance object of trade, possession, etc	String		Regular Expression
absolute_weight	Absolute weight of the illegal substance	Float	Grams	Regular Expression
doses	Number of doses of illegal substance	Float		Regular Expression
weight_active_principle	Weight of the illegal substance relative to the absolute number	Float	Grams	Regular Expression
percentage	Percentage of purity of the substance	Float	Percentage	Regular Expression
punishment_days	Days of confinement in jail	Integer	Days	Regular Expression
pecuniary_punishment_amount	Amount of the pecuniary sanction	Integer	Euro	Regular Expression
minor_offence	Whether the offence has been considered non-minor (0) or minor (1)	Boolean		Manual
recidivism	Whether the offender is not a re-offender (0) or he/she is a re-offender (1)	Boolean		Manual
plea_bargain	Whether the offender has not negotiated a plea bargain (0) or he/she has (1)	Boolean		Manual
shortened_proceeding	Whether the offender has opted for full proceeding (0) a shortened proceeding (1)	Boolean		Manual

Some clarifications on certain information types might be necessary. Regarding **substance**, it is worth noticing that judges usually prefer the common name (e.g., cocaine, hashish) instead of the scientific name reported provided by the law. Therefore, the regular expressions used to extract the information did not include scientific formulae in favour of the common name. In the case multiple substances have been found in the same case, the one with the highest number of doses have been kept. Regarding **absolute_weight** and **weight_active_principle**, the regular expressions contemplate several units of measures ("kg", "mg", "mgr", etc.), also in full-text word in Italian (e.g., "chilogrammo").

The extracted information have been standardised in grams, which was the most common unit of measure. Regarding `percentage` and `weight_active_principle`, missing data have been calculated by dividing or multiplying the available information types according to the target data. Regarding `punishment_days`, the regular expression matches all the possible combinations of year, month, and days expressed in numeric or string format. Dictionary-based conversions were used to convert string to numbers, and then multiplications to standardise the duration of the punishments in days were done. Regarding `pecuniary_punishment_amount`, the extracted information consisted of numbers, string and combinations of the two. As with the previous case, dictionary-based conversions were used to make the amount unique. The other information types have been extracted manually from the text of the ruling. Regarding `substance`, a closed list of the most frequent illegal narcotics (marijuana, heroine, cocaine, hashish, crack, ketamine) have been encoded to be represented as a categorical attribute. Finally, regarding Regarding `doses`, the number has been either extracted by the ruling or calculated by applying the average weight : dose ratio available from cases in which the same substance was traded.

6   Findings from Data Analysis

Following the hybrid methodology described in Sect. 3, an unsupervised K-Means clustering (K = 3) experiment was run to understand whether groups of "minor" and "non-minor" rulings can be automatically identified taking into account all the available attributes, with potential outliers a third option for outliers. The results are displayed in Fig. 2 with days of punishment and amount of pecuniary sanction as axes.

As expected, a division can be observed between a first groups of rulings (highlighted in red), possibly associated to "minor" cases, and another group of rulings (highlighted in green). A third group was added to the clustering to identify potential outliers. This group likely corresponds to crimes in which the case has been qualified as a drug dealing action in which a "huge amount" of narcotics to art. 80(2) of the Presidential Decree 309/1990. This legal qualification is not the object of this study. However, this additional legal factor identified in the unsupervised experiment signals that there is potential to expand the study in this direction. For the purposes of this study, the division between "minor" and "non-minor" seems present in the features of the dataset. Therefore, the analysis can go on without the necessity of immediately refining the light taxonomy and proceeds with additional features. First, we run an experiment to determine what model is more promising keeping constant certain parameters. We use `scikit-learn` in Python[8] instantiating a Decision Tree and a Random

---

[8] Scikit-learn is a Python library for machine learning, providing tools for data pre-processing, classification, regression, clustering, and more.

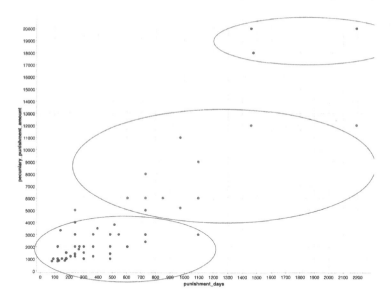

Fig. 2. Clustering of cases with K=3. The chart displays groups of court rulings (red and green) for "minor" and "non-minor" offences. A third group (brown), contains outliers which might include cases in which the aggravating circumstance for "serious" offences has been applied (Color figure online)

Forest models with `max_depth=8`. At the same time, similar experiments were run with KNIME[9].

The table presents the results of several classification experiments using different machine learning algorithms, namely Decision Tree, Support Vector Machine (SVM), Random Forest, and Gradient Boosted Trees. The classification tasks uses the predictive features to classify `minor_offense`. The experiments were performed with varying training and test ratios of 80:20, 75:25, and 70:30 to evaluate the impact of dataset compositions in a binary classification with few data available. The evaluation metrics used to compare the algorithms were precision, accuracy, recall, and F1-measure[10].

Observing the results, it can be seen that the performance of the algorithms varies significantly depending on the training and test ratios. In general, the Decision Tree algorithm performed well, especially when using the KNIME

---

[9] KNIME is a data analytics platform that allows users to visually design workflows, integrating various data processing and machine learning algorithms for advanced analytics and predictive modelling.

[10] These metrics are widely utilized for assessing the performance of classification models. Precision and recall are especially valuable for datasets that exhibit significant class imbalance, wherein the number of instances in various classes differs substantially. In contrast, accuracy is well-suited for balanced datasets. The F1-measure is a composite metric that integrates both precision and recall and is considered a more equitable evaluation criterion for classification models.

Table 2. This table presents classification metrics for different models. The metrics include precision, accuracy, recall, and F1-measure, calculated for different ratios of training to test data.

Model	Training: Test	Precision	Accuracy	Recall	F1-measure
Decision Tree (scikit-learn)	**80:20**	**0.55**	**0.56**	**0.56**	**0.55**
Decision Tree (scikit-learn)	75:25	0.50	0.50	0.50	0.49
Decision Tree (scikit-learn)	70:30	0.45	0.44	0.45	0.44
SVM (scikit-learn)	80:20	0.38	0.39	0.39	0.37
SVM (scikit-learn)	75:25	0.34	0.36	0.36	0.34
SVM (scikit-learn)	70:30	0.42	0.44	0.44	0.42
Random Forest (scikit-learn)	80:20	0.34	0.39	0.39	0.34
Random Forest (scikit-learn)	75:25	0.39	0.41	0.41	0.38
Random Forest (scikit-learn)	70:30	0.48	0.48	0.48	0.47
Decision Tree (KNIME)	80:20	0.61	0.61	0.63	0.61
Decision Tree (KNIME)	75:25	0.53	0.55	0.53	0.51
Decision Tree (KNIME)	**70:30**	**0.65**	**0.65**	**0.65**	**0.65**
Tree Ensemble (KNIME)	80:20	0.54	0.53	0.56	0.53
Tree Ensemble (KNIME)	75:25	0.64	0.54	0.54	0.54
Tree Ensemble (KNIME)	70:30	0.54	0.62	0.53	0.49
Random Forest (KNIME)	80:20	0.63	0.61	0.58	0.54
Random Forest (KNIME)	75:25	0.64	0.54	0.54	0.54
Random Forest (KNIME)	70:30	0.60	0.58	0.58	0.54
Gradient Boosted Trees (KNIME)	80:20	0.61	0.61	0.61	0.61
Gradient Boosted Trees (KNIME)	75:25	0.59	0.55	0.56	0.55
Gradient Boosted Trees (KNIME)	70:30	0.54	0.54	0.54	0.54

implementation, achieving higher precision, accuracy, recall, and F1-measure across all training and test ratios. The SVM algorithm, on the other hand, performed poorly, with relatively low precision, accuracy, recall, and F1-measure across all ratios. The Random Forest algorithm achieved moderate to good results, with the KNIME implementation generally performing better than the scikit-learn implementation (Table 2).

After evaluating the performance of the different machine learning algorithms, it is worth exploring some of the factors that may explain the observed differences between the two sets of algorithms. With this regard, they may be explained by several factors. For instance, the decision tree algorithm used in KNIME may have a different stopping criterion than the one used in scikit-learn, which can result in different tree structures and ultimately, different classification accuracy. Also, Scikit-learn and KNIME may have different default hyperparameters for the same algorithm.

The choice of hyperparameters can have a significant impact on the performance of the algorithm, and if the hyperparameters are not tuned properly, it can lead to suboptimal performance. In these experiments, hyperparameters were left untouched. `GridSearchCV` method from Sckit-learn was used to perform an exhaustive search over specified hyperparameter values for a given estimator. This method performs cross-validation for each combination of hyperparameters and returns the best set of hyperparameters that yields the highest score on a specified evaluation metric. The improved model had a theoretical F1-measure of 0.67, which is compatible to the KNIME one. Since the decision tree was the best performing algorithm in both tests, it is convenient to visualise the criteria on which the classification has been performed. The visualisations purposively display the most influential factors from both algorithms to ease the global understanding and interpretability of the results (Figs. 3 and 4).

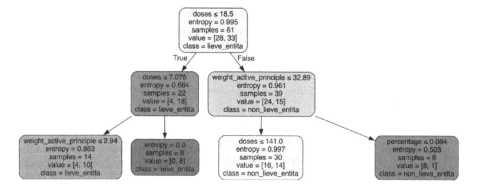

Fig. 3. The best-performing decision tree using Sckit-learn. Blue-highlighted cells represent the relevance for "minor" classification, orange colours stand for relevance of the factor in "non-minor" classifications. Darker colours represent higher relevance (Color figure online)

In both cases, the first criterion is `doses`, which varies from <18.5 and >18.5 (Sckit-learn) and <17.5 and >17.5 (KNIME) for "minor" and "non-minor" classification respectively. The difference may be due to the same factors (hyperparameters) identified when discussing the discrepancy in performance. However, they seem highly compatible given the small quantitative differences among them. In both cases, the quantity of `weight_active_principle` seems quite relevant as it is ranked among the most influential factors. As it will be discussed in the next section, this is expected with the legal analysis carried out in the preliminary stages of the study.

With doses being ranked as the most influential factor in the two best-performing experiments, further visualisations allow the interpretability of the results. Figure 5 below. The position of the rulings based some extracted factors such as `doses` can be visualized in 3 dimensions to be correlated to `pecuniary_punishment_amount`, and `punishment_days`, which should express the "minor" or "non-minor" punishment in tangible terms. Additionally, the volume of dots should to `weight_active_principle` which was considered a relevant factor alongside `doses`. Green dots represent "minor" cases, whereas red dots display "non-minor" cases.

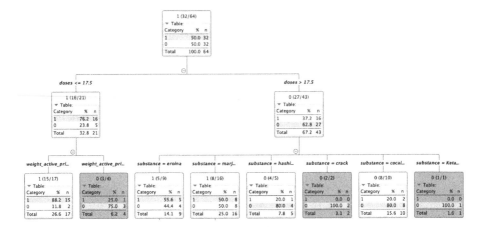

Fig. 4. The best-performing decision tree using KNIME. Each box contains and indication of the criterion and the number of instances from the training set (N = 64) classified according to the factor in bod (e.g. doses). Orange boxes are terminal leaves (Color figure online)

The figure seems to confirm the expected distribution of cases. "Minor" cases are close to the lower number of doses and receive a lower amount of punishment, both in pecuniary and sentencing terms. Outliers are represented by cases that, despite a lower number of doses, display higher punishments. Such cases seem related to the higher number of `weight_active_principle`, which was ranked as a relevant factor in the previous steps. This indicates a higher level of danger despite the lower number of doses, which might be linked to the purity of the traded narcotics.

7   Legal Analysis and Limitations of the Results

The analysis performed over a legal corpus allows for a legal interpretation of the findings. First, the research question posed in Sect. 3 might be answered by proposing the number of doses as the most important factor in the legal qualification of micro-trafficking facts, with the weight of the active principle playing a relevant role. This is confirmed by the double validation performed with different sets of algorithms. Despite some relatively small differences, the two approaches came to the same conclusion under different testing scenarios, in particular as regards the ratio between training and test set.

This finding seems to be confirmed by the recent indications of the Italian Court of Cassation, which has identified some thresholds in the amount (expressed in weight) that might help courts of merits in the legal qualification of the facts. While this jurisprudential trend is on its way of consolidation (in the absence of clear legislative indications), this analysis also contributes to this debate by providing a different perspective on a purely legal matter, which identifies the number of doses as the most influential criterion in courts of merits.

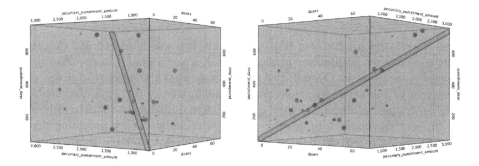

Fig. 5. 3-dimensional visualisations from two different angles of the position the rulings according to `doses`, `pecuniary_punishment_amount`, and `punishment_days`. Dots' volume represents `weight_active_principle`. The green plane represents an ideal linear trend in which the variables are directly proportionate

This paper does not explore further this divergence, as an in-depth discussion would require a different methodology aimed at investigating, as shown in other studies (e.g., in [20]) pros and cons of one quantitative criterion over the other.

Overall, the approach used in this study favours the explainability of the machine learning systems developed to support the legal analysis. By understanding, also by means of visualisation, what the impact of each feature in the final decision, a reasonable margin of appreciation is left to the legal expert in interpreting the results from a legal standpoint.

Let us also address some important limitations of this study. First and most importantly, the dataset used to perform the analysis is relatively small and it has to be increased in further refinements. There are important factors to keep into account. For instance, compliance with data protection law and privacy is a key aspect which has to be investigated before introducing new elements to the dataset. Another limitation regards the quantity and the quality of the feature extracted from the rulings. One key element to underline is that the light taxonomy developed in the early stages of the methodology constitutes an "open world" prone to be enriched by additional elements. Criminal rulings in microtrafficking present several factual factors that should be kept into account in further refinements (e.g., precautionary measures). At the same time, increasing the complexity of the knowledge on which the model finds correlations and patterns may hinder the overall performances.

This paper has found a trade-off by relying on a method capable of grasping the *legal* significance of pre-identified factors. There are potential avenues for exploring other methods. One could be the automated extraction of linguistic formulas and patterns, already tested in other studies in the legal domain [13]. In this perspective study, it would be necessary to reconcile the linguistic patterns to their relevance in the legal language. Ontologies seem a viable way to perform such bottom-up approach in combination with the top-down pre-identification of the legal elements performed in this study.

8   Conclusive Remarks

Given the relative small size of this legal corpus, it is worth underlining that this study constitutes a preliminary attempt of analysing criminal rulings rather than a complete assessment. Overall, this study shows that the number of doses and the weight of the active principle are relevant factors in the legal qualification of micro-trafficking criminal cases in the Italian jurisdiction. This conclusion can support the research of scholars active in criminal law by providing novel perspectives on a current debate. Although significant under a legal point of view, much has to be done to ensure the reliability of the findings of this paper, in particular in the area of data availability. This might affect the generalizability of the results.

However, there are some important takeaways that deserve attention. First, drug micro-trafficking is a legal sector that can be explored with the lens of legal informatics for further experiments. Future studies should increase the "datafication" of rulings by expanding the factors to be taken into account, in particular in the direction of qualitative factual elements such as indications and evidence. Second, the methodology used in this paper seems appropriate for similar tasks and can be further explored in the domain of legal informatics. Finally, explainability and interpretability of results are necessary to validate the outputs of machine learning algorithms.

Acknowledgements. This presentation has been funded by Programma Operativo Nazionale (PON) "Governance e Capacità Istituzionale" 2014–2020, "Universitas per la Giustizia. Programma per la qualità del sistema giustizia e per l'effettività del giusto processo (UNI 4 JUSTICE) - Macro Area 2 - CUP J19J21026980006"

References

1. Verheij, B.: Artificial intelligence as law. Artif. Intell. Law **28**(2), 181–206 (2020)
2. Ashley, K.D.: Artificial Intelligence and Legal Analytics: New Tools for Law Practice in the Digital Age. Cambridge University Press, Cambridge (2017)
3. Athan, T., Governatori, G., Palmirani, M., Paschke, A., Wyner, A.: LegalRuleML: design principles and foundations. In: Faber, W., Paschke, A. (eds.) Reasoning Web 2015. LNCS, vol. 9203, pp. 151–188. Springer, Cham (2015). https://doi.org/10.1007/978-3-319-21768-0_6
4. Prakken, H., Sartor, G.: Modelling reasoning with precedents in a formal dialogue game. In: Judicial Applications of Artificial Intelligence, pp. 127–183. Springer, Heidelberg (1998). https://doi.org/10.1007/978-94-015-9010-5_5
5. Prakken, H., Sartor, G.: Law and logic: a review from an argumentation perspective. Artif. Intell. **227**, 214–245 (2015)
6. Liga, D., Palmirani, M.: Classifying argumentative stances of opposition using tree kernels. In: Proceedings of the 2019 2nd International Conference on Algorithms, Computing and Artificial Intelligence, pp. 17–22 (2019)
7. Wyner, A., Peters, W.: On rule extraction from regulations. In: Legal Knowledge and Information Systems, pp. 113–122. IOS Press (2011)

8. Palmirani, M.: Legislative change management with Akoma-Ntoso. In: Legislative XML for the Semantic Web, pp. 101–130. Springer, Heidelberg (2011). https://doi.org/10.1007/978-94-007-1887-6_7

9. Sartor, G., et al.: Legislative XML for the Semantic Web: Principles, Models, Standards for Document Management, vol. 4. Springer, Heidelberg (2011). https://doi.org/10.1007/978-94-007-1887-6

10. Ashley, K.D.: Automatically extracting meaning from legal texts: opportunities and challenges. Ga. St. UL Rev. **35**, 1117 (2018)

11. Robaldo, L., et al.: Introduction for artificial intelligence and law: special issue "natural language processing for legal texts" (2019)

12. Palmirani, M., et al.: Legal ontology for modelling GDPR concepts and norms. In: Legal Knowledge and Information Systems, pp. 91–100. IOS Press (2018)

13. Palmirani, M., et al.: PrOnto ontology refinement through open knowledge extraction. In: Legal Knowledge and Information Systems, pp. 205–210. IOS Press (2019)

14. Palmirani, M., et al.: Hybrid AI framework for legal analysis of the EU legislation corrigenda. In: Legal Knowledge and Information Systems, pp. 68–75. IOS Press (2021)

15. Palmirani, M., Sapienza, S.: Big data, explanations and knowability. Ragion pratica **2**, 349–364 (2021)

16. Atkinson, K., Bench-Capon, T., Bollegala, D.: Explanation in AI and law: past, present and future. Artif. Intell. **289**, 103387 (2020)

17. Sapienza, S., Bomprezzi, C., et al.: Algorithmic Justice e classificazione di rischio nella Proposta AI ACT. In: La Trasformazione Digitale della Giustizia nel dialogo tra discipline: Diritto e Intelligenza Artificiale, pp. 65–114. Giuffré Francis Lefebvre (2022)

18. Gambardella, M.: Illeciti in materia di stupefacenti e riforma dei reati contro la persona: un antidoto contro le sostanze velenose. Cassazione penale **60**(2), 449–486 (2020)

19. Hall, J.: Nulla poena sine lege. Yale Law J. **47**(2), 165–193 (1937)

20. Lombardi, F.: La Cassazione fissa le soglie quantitative per la lieve entità ex art. 73 co. 5 DPR 309/1990. Osservazioni a prima lettura. In: Giurisprudenza Penale Web, vol. 12 (2022)

21. Palmirani, M., Vitali, F.: Akoma-Ntoso for legal documents. In: Legislative XML for the semantic Web, pp. 75–100. Springer, Heidelberg (2011). https://doi.org/10.1007/978-94-007-1887-6_6

22. Ceci, M., Palmirani, M.: Ontology framework for judgment modelling. In: Palmirani, M., Pagallo, U., Casanovas, P., Sartor, G. (eds.) AICOL 2011. LNCS (LNAI), vol. 7639, pp. 116–130. Springer, Heidelberg (2012). https://doi.org/10.1007/978-3-642-35731-2_8

Exploring the Potential of AI&MDL
for Enhancing E-Government Services:
A Review Paper

Asefeh Asemi[1] , Adeleh Asemi[2] , and Andrea Ko[1(✉)]

[1] Corvinus University of Budapest, Fovam Ter 1093, Budapest, Hungary
andrea.ko@uni-corvinus.hu
[2] Faculty of Computer Science and Information Technology, Universiti Malaya,
50603 Kuala Lumpur, Malaysia

Abstract. E-government services aim to make government services more accessible, efficient, and user-friendly through electronic means. The combination of artificial intelligence, machine learning, and deep learning (AI&MDL) can help to automate manual processes, reduce waiting time, and provide more personalized services. This study explores the potential of AI&MDL for enhancing e-government services through a comprehensive review of the available literature. The study focuses on three key research questions: (1) How can AI&MDL be used in the efficiency of e-government services? (2) What are the potential benefits and challenges associated with the implementation of AI&MDL in e-government services? (3) What strategies can be employed to ensure the security and privacy of data collected through AI&MDL in e-government services? The literature review reveals that AI&MDL can greatly use for the efficiency of e-government services and provide several benefits, including improved service quality, increased transparency, and enhanced citizen engagement. However, implementing AI&MDL also presents challenges, such as security and privacy risks and the need for significant investments in technology and infrastructure. Governments must consider the security and privacy of sensitive data and implement strategies to ensure that AI&MDL systems are transparent, accountable, and secure. In conclusion, AI&MDL has the potential to revolutionize e-government services, but the implementation must be carefully considered to ensure that the benefits are maximized while minimizing the potential risks. This study provides valuable insights into the current state and offers important considerations for governments and policymakers.

Keywords: Artificial Intelligence · Machine Learning · Deep Learning · E-Government Services · AI&MDL

1 Introduction

Artificial Intelligence (AI) and Machine/Deep Learning (MDL) have revolutionized many industries and are now being applied to e-government services. E-government services play a crucial role in providing citizens with access to essential services, and the

A. Kö et al. (Eds.): EGOVIS 2023, LNCS 14149, pp. 118–138, 2023.
https://doi.org/10.1007/978-3-031-39841-4_9

combination of them has the potential to enhance their efficiency and effectiveness significantly [1]. Artificial Intelligence refers to developing computer systems that can perform tasks that require human intelligence, such as visual perception, speech recognition, decision-making, and language translation [2]. AI is a broad field that encompasses many different subfields, including machine learning, natural language processing, robotics, and computer vision. The goal of AI research is to create intelligent machines that can perform tasks that would otherwise require human intervention, with the potential to revolutionize many different areas of our lives. ML is a subset of AI that focuses on developing algorithms that enable computers to learn from data, identify patterns, and make predictions [3]. DL, which is a type of machine learning, uses artificial neural networks to model complex relationships between inputs and outputs. This enables deep learning algorithms to automatically improve their performance on tasks such as image and speech recognition, without being explicitly programmed. E-government services refer to the delivery of government services through electronic means, such as websites, portals, or mobile applications [4]. The goal of e-government services is to make government services more accessible, efficient, and user-friendly, reducing the need for citizens to visit government offices in person [5]. By leveraging these techniques, e-government services can provide more personalized and efficient services and improve the overall experience for citizens. This study explores the potential of these techniques for enhancing e-government services through a comprehensive review of the available literature. In this paper, intelligent e-government systems refer to the development of systems that can perform tasks that would typically require human intelligence to complete. These systems use algorithms and statistical models to recognize patterns and make predictions based on data. In these systems, ML considers a subset of them that involves training algorithms to make predictions or decisions based on data. DL considers a subset of ML that uses neural networks with many layers to learn from large amounts of data. Our study proposed a combination of AI&MDL applications to analyze large datasets and make predictions in e-government. The study focused on three key research questions: (1) How can AI&MDL be used in the efficiency of e-government services? (2) What are the potential benefits and challenges associated with the implementation of AI&MDL in e-government services? (3) What strategies can be employed to ensure the security and privacy of data collected through AI&MDL in e-government services? The study is based on a review of the literature and library studies, providing valuable insights into the current state of AI&MDL in e-government services. This study provides a comprehensive overview of the opportunities and challenges of AI&MDL in the context of e-government services and offers important considerations for governments and policymakers.

2 Methods

This study is a qualitative review that examined documents related to AI&MDL and electronic government. The study's statistical population comprises relevant documents indexed in Web of Science, Scopus, and PubMed databases. Table 1 summarizes the methods, techniques, and tools/software suitable for each step. The research methodology was carried out systematically as the following:

1. Search for relevant literature in online databases such as the Web of Science (WoS).
2. Identify keywords and phrases related to your topic and use them to search for relevant articles.
3. Read the abstracts of the articles to determine if they are relevant to your research question or topic.
4. Read the full text of the articles relevant to your research question or topic and take notes on important information.
5. Analyze the information you have gathered from the literature review and identify any gaps in knowledge or areas that need further exploration.
6. Synthesize the information you have gathered from the literature review into a cohesive argument or narrative about your research question or topic.
7. Cite all sources used in your literature review using a standard citation format such as APA.

The implementation research: On March 3rd, 2023, a search was conducted on three academic databases to retrieve documents related to the intersection of artificial intelligence (Topic), machine learning or deep learning (Topic), and e-government services (Topic). A search in Web of Science (WoS) using "Artificial Intelligence" and ("Machine Learning" OR "Deep Learning") and "E-Government" yielded 20 documents. A search in Scopus using "TITLE-ABS-KEY (artificial AND intelligence) AND TITLE-ABS-KEY (machine AND learning OR deep AND learning) AND TITLE-ABS-KEY (e-government AND services)" yielded 24 documents. No documents were retrieved from PubMed using the search terms "artificial intelligence" AND ("machine learning" OR "deep learning") AND ("e-government services"). A total of 44 full records were imported into Zotero, and after removing nine duplicate records, 35 documents were retained. To identify relevant articles, a list of keywords and phrases related to the topic was generated by brainstorming and using synonyms and related terms. Thesauruses and keyword suggestion tools such as Google Trends and LC Subject Heading were also used to expand the list of terms. Table 2 displays the list of synonyms or related terms for each phrase.

Using the Web of Science database and the search query "Artificial Intelligence OR AI OR Cognitive computing OR Intelligent automation OR Machine intelligence OR Synthetic intelligence (Topic) and Machine Learning OR ML OR Supervised learning OR Unsupervised learning OR Reinforcement learning OR Decision tree learning OR Neural networks OR Deep Learning OR DL OR Deep neural networks OR Convolutional neural networks OR Recurrent neural networks OR Autoencoders OR Generative adversarial networks (Topic) and E-Government Services OR Electronic government services OR Digital government services OR Online government services OR govtech OR Smart government services OR Cyber government services (Topic)", a total of 63

Table 1. Research methods, techniques, and tools/software

Step	Method	Technique	Tools/Software
1. Search for relevant literature in online databases	A systematic search using specific keywords and Boolean operators (AND, OR, NOT)	Use truncation, wildcards, and proximity operators to broaden or narrow the search. Use filters such as publication date, study design, and language to limit the search results Search Formula in WoS: Artificial Intelligence (Topic) and Machine Learning OR Deep Learning (Topic) and E-Government (Topic) Search Formula in Scopus: (TITLE-ABS-KEY (artificial AND intelligence) AND TITLE-ABS-KEY (machine AND learning OR deep AND learning) AND TITLE-ABS-KEY (e-government AND services)) Search Formula in PubMed: (artificial AND intelligence)) AND (machine learning OR deep learning)) AND (e-government services)	Web of Science, Scopus, PubMed
2. Identify keywords and phrases related to your topic and use them to search for relevant articles	Brainstorm and create a list of keywords and phrases related to your topic	Use synonyms and related terms for your keywords to expand the search results. Use a thesaurus or keyword suggestion tool to find additional terms	Google Trends, LC Subject Heading
3. Read the abstracts of the articles to determine if they are relevant to your research question or topic	Read the abstracts of articles identified in the search	Use specific criteria or questions to evaluate the relevance of the abstract	N/A

(*continued*)

Table 1. (*continued*)

Step	Method	Technique	Tools/Software
4. Read the full text of the articles that are relevant to your research question or topic and take notes on important information	Read the full text of relevant articles. Take notes on important information	Use a systematic approach to reading the articles, such as scanning the article first to identify key sections and then reading more closely. Use a note-taking tool or software to organize and summarize the information	Zotero
5. Analyze the information you have gathered from the literature review and identify any gaps in knowledge or areas that need further exploration	Analyze the information gathered from the literature review	Use a systematic approach to analyzing the information, such as categorizing the information into themes or concepts. Identify patterns or inconsistencies in the information	N/A
6. Synthesize the information you have gathered from the literature review into a cohesive argument or narrative about your research question or topic	Synthesize the information gathered from the literature review	Use a systematic approach to synthesizing the information, such as identifying key themes or concepts and organizing the information into a logical structure	N/A

documents were retrieved. In the Scopus database, acronyms were removed due to the restriction of using abbreviations in field searches. The following search query was used: "(ALL (artificial AND intelligence OR cognitive AND computing OR intelligent AND automation OR machine AND intelligence OR synthetic AND intelligence) AND ALL (machine AND learning OR supervised AND learning OR unsupervised AND learning OR reinforcement AND learning OR decision AND tree AND learning OR neural AND networks OR deep AND learning OR deep AND neural AND networks OR recurrent AND neural AND networks OR generative AND adversarial AND networks) AND ALL (e-government AND services OR electronic AND government AND services OR digital AND government AND services OR online AND government AND services OR smart AND government AND services OR cyber AND government AND services))". A total of 14 documents were retrieved.

Table 2. The list of synonyms or related terms

Artificial Intelligence	Machine Learning	Deep Learning	E-Government Services
AI	ML	DL	Electronic government services
Cognitive computing	Supervised learning	Deep neural networks	Digital government services
Intelligent automation	Unsupervised learning	Convolutional neural networks	Online government services
Machine intelligence	Reinforcement learning	Recurrent neural networks	GovTech
Synthetic intelligence	Decision tree learning	Autoencoders	Smart government services
Neural networks		Generative adversarial networks	Cyber government services

3 Finding

After combining all results and removing duplicates, finally, after reading the abstracts, 84 documents remained including 47 journal papers and 30 conference papers, 5 conference books, and 2 book chapters. Conference books were removed from the list because of duplicate papers.

The Table 3 consists of various items related to e-government and artificial intelligence such as journal articles, conference papers, and book sections. The table contains information regarding the publication year, author, and title of the item. Upon analyzing the Table 3, it can be observed that the topics covered are diverse, including the use of AI in e-government, digital transformation impact on society [6, 7], smart and sustainable cities [8–19], and intelligent cyber threat identification in smart city environments [17, 20–22]. There is a significant focus on the use of machine learning and AI in e-government [12, 17, 22–25, 30–33], including topics such as applying explainable AI techniques, establishing efficient governance through data-driven e-government, and transforming e-participation. Furthermore, the table includes several studies and reviews on the use of AI&MDL in various sectors such as education [35], public sector [12, 34], [35] and the prediction of firms' vulnerability to economic crises [21, 25–27]. The table also includes studies on chatbots [28, 29], risk management frameworks, and the impact of decision-makers' experience on AI-supported decision-making in government. Overall, the table provides a comprehensive overview of the various topics and studies related to e-government and AI.

Table 3. Resources related to e-government and artificial intelligence

Item Type	Year	Author	Title
Conference	2007	Chun	Using AI for e-Government automatic assessment of immigration application forms
Conference	2008	Chun	An AI framework for the automatic assessment of e-government forms
Conference	2015	Said et al.	Exploiting Computational intelligence Paradigms in e-Technologies and Activities
Journal	2015	Yu et al.	Social credit: a comprehensive literature review
Book Chapter	2016	Coria	Open data for human development in Mexico: A data science perspective
Journal	2016	Zhang	Humans and Machines in the Evolution of AI in Korea
Conference	2017	Xue et al.	A New Governance Architecture for Government Information Resources based on Big Data Ecological Environment in China
Conference	2017	Grandhi et al.	Data driven marketing for growth and profitability
Conference	2018	Park et al.	How to Define Value on Data under Blockchain Driven Open Data System for E-Government
Journal	2018	Zoltán & Jinil	Taxonomy, use cases, strengths and challenges of chatbots
Conference	2018	Agbozo & Spassov	Establishing efficient governance through data-driven e-government
Conference	2018	Manocha et al.	Technological trends, impact and analysis of social media quality parameters on e-governance applications
Journal	2018	Caruso	Digital innovation and the fourth industrial revolution: epochal social changes?
Journal	2018	Ju et al.	Citizen-centered big data analysis-driven governance intelligence framework for smart cities

(continued)

Table 3. (*continued*)

Item Type	Year	Author	Title
Conference	2019	Lopes et al.	Pannel: Digital Transformation Impact on Society
Conference	2019	Ronzhyn et al.	Scenario technique to elicit research and training needs in digital government employing disruptive technologies
Conference	2019	Alexopoulos et al.	How machine learning is changing e-government
Journal	2019	Al-Mushayt	Automating E-Government Services with Artificial Intelligence
Journal	2019	de Sousa	How and where is artificial intelligence in the public sector going? A literature review and research agenda
Conference	2019	Dominguez et al.	Privacy and information protection for a new generation of city services
Conference	2019	Vuppalapati et al.	The Development of Machine Learning Infused Outpatient Prognostic Models for tackling Impacts of Climate Change and ensuring Delivery of Effective Population Health Services
Journal	2019	Soomro et al.	Smart city big data analytics: An advanced review
Conference	2019	Jin et al.	Diagnosis of Corporate Insolvency using Massive News Articles for Credit Management
Journal	2020	Bannister & Connolly	Administration by algorithm: A risk management framework
Book Chapter	2020	Narasiman et al.	IndQuery - An Online Portal for Registering E-Complaints Integrated with Smart Chatbot
Journal	2020	Yfantis et al.	Exploring the Implementation of Artificial Intelligence in the Public Sector: Welcome to the Clerkless Public Offices. Applications in Education

(*continued*)

Table 3. (*continued*)

Item Type	Year	Author	Title
Conference	2020	Loukis et al.	Using Government Data and Machine Learning for Predicting Firms' Vulnerability to Economic Crisis
Journal	2020	Lathrop	The Inadequacies of the Cybersecurity Information Sharing Act of 2015 in the Age of Artificial Intelligence
Journal	2020	Henman	Improving public services using artificial intelligence: possibilities, pitfalls, governance
Journal	2020	Alanazi et al.	Measuring and Preventing COVID-19 Using the SIR Model and Machine Learning in Smart Health Care
Journal	2020	Hausken	Cyber resilience in firms, organizations and societies
Conference	2020	Niture et al.	AI Based Airplane Air Pollution Identification Architecture Using Satellite Imagery
Journal	2020	Tkatek et al.	Putting the world back to work: An expert system using big data and artificial intelligence in combating the spread of COVID-19 and similar contagious diseases
Conference	2020	Jayanthi et al.	Fourth Industrial Revolution: An Impact on Health Care Industry
Conference	2020	Patnaik et al.	Use of Data Analytics for Effective E-Governance: A Case Study of "EMutation" System of Odisha
Conference	2020	Shadbolt & Assoc Comp Machinery	Architectures for Autonomy: Towards an Equitable Web of Data in the Age of AI
Conference	2020	Likhacheva et al.	Use of Digital Technologies by Financial Services of Russian Companies
Journal	2020	Clavería	General Practice and the Community: Research on health service, quality improvements and training. Selected abstracts from the EGPRN Meeting in Vigo

(*continued*)

Table 3. (*continued*)

Item Type	Year	Author	Title
Conference	2021	Owoc & Weichbroth	University Students' Research on Artificial Intelligence and Knowledge Management. A Review and Report of Multi-case Studies
Conference	2021	Kalampokis et al.	Applying Explainable Artificial Intelligence Techniques on Linked Open Government Data
Conference	2021	Dreyling et al.	Social, Legal, and Technical Considerations for Machine Learning and Artificial Intelligence Systems in Government
Journal	2021	Hujran et al.	Digitally Transforming Electronic Governments into Smart Governments: SMARTGOV, an Extended Maturity Model
Journal	2021	Zhang et al.	Big data and artificial intelligence based early risk warning system of fire hazard for smart cities
Journal	2021	Zekic-Susac et al.	Machine learning based system for managing energy efficiency of public sector as an approach towards smart cities
Journal	2021	Basit et al.	A comprehensive survey of AI-enabled phishing attacks detection techniques
Conference	2021	Peterlin et al.	Engineering Technology-Based Social Innovations Accommodating Functional Decline of Older Adults
Journal	2021	Firouzi et al.	Harnessing the Power of Smart and Connected Health to Tackle COVID-19: IoT, AI, Robotics, and Blockchain for a Better World
Journal	2021	Yuan & Elhoseny	Artificial Intelligence as a maturing and growing technology: An urgent need for intelligent systems
Journal	2021	Abid et al.	Toward an Integrated Disaster Management Approach: How Artificial Intelligence Can Boost Disaster Management

(*continued*)

Table 3. (*continued*)

Item Type	Year	Author	Title
Journal	2021	Deng et al.	Research on Convolutional Neural Network-Based Virtual Reality Platform Framework for the Intangible Cultural Heritage Conservation of China Hainan Li Nationality: Boat-Shaped House as an Example
Journal	2021	Rawindaran et al.	Machine Learning Cybersecurity Adoption in Small and Medium Enterprises in Developed Countries
Journal	2021	Cavus et al.	An Artificial Intelligence-Based Model for Prediction of Parameters Affecting Sustainable Growth of Mobile Banking Apps
Journal	2021	Bozorgastl et al.	New Trends on Digital Twin-Based Blockchain Technology in Zero-Emission Ship Applications
Journal	2021	Ozbay & Dalcali	Effects of COVID-19 on electric energy consumption in Turkey and ANN-based short-term forecasting
Journal	2021	Bharanidharan & Jayalakshmi	Predictive Scaling for Elastic Compute Resources on Public Cloud Utilizing Deep Learning based Long Short-term Memory
Journal	2021	Alwabel & Zeng	Data-driven modeling of technology acceptance: A machine learning perspective
Journal	2021	Zhang et al.	Deep-Learning-Empowered 3D Reconstruction for Dehazed Images in IoT-Enhanced Smart Cities
Journal	2021	Kaginalkar et al.	Review of urban computing in air quality management as smart city service: An integrated IoT, AI, and cloud technology perspective
Journal	2021	Bereznoi	Global oil & gas corporations in the race for technological superiority
Journal	2021	Khowaja et al.	VIRFIM: an AI and Internet of Medical Things-driven framework for healthcare using smart sensors

(*continued*)

Table 3. (*continued*)

Item Type	Year	Author	Title
Journal	2021	Auboin & Koopman	Trade and innovation policies: Coexistence and spillovers
Journal	2022	Espina-Romero & Guerrero-Alcedo	Fields Touched by Digitalization: Analysis of Scientific Activity in Scopus
Conference	2022	Bansal et al.	Advancing e-Government Using Internet of Things
Journal	2022	Janssen et al.	Will Algorithms Blind People? The Effect of Explainable AI and Decision-Makers' Experience on AI-supported Decision-Making in Government
Journal	2022	Porwol et al.	Transforming e-participation: VR-dialogue - building and evaluating an AI-supported framework for next-gen VR-enabled e-participation research
Journal	2022	Strielkowski et al.	Management of Smart and Sustainable Cities in the Post-COVID-19 Era: Lessons and Implications
Conference	2022	Kenana	Should the Governments Promote or Control Development in Machine Learning and Artificial Intelligence AI?
Conference	2022	Mukherjee	Artificial Intelligence Based Smart Government Enterprise Architecture (AI-SGEA) Framework
Journal	2022	Al-Taleb & Saqib	Towards a Hybrid Machine Learning Model for Intelligent Cyber Threat Identification in Smart City Environments
Journal	2022	Wang & Wang	Deep Learning Models and Social Governance Guided by Fair Policies
Journal	2022	Guberovic et al.	Framework for federated learning open models in e-government applications

(*continued*)

Table 3. (*continued*)

Item Type	Year	Author	Title
Journal	2022	Hilal et al.	Artificial Intelligence Based Sentiment Analysis for Health Crisis Management in Smart Cities
Conference	2022	Bennett & Robertson	An AI-based framework for remote sensing supporting multi-domain operations
Journal	2022	Valle-Cruz et al.	From E-budgeting to smart budgeting: Exploring the potential of artificial intelligence in government decision-making for resource allocation
Journal	2022	Alahmari et al.	Musawah: A Data-Driven AI Approach and Tool to Co-Create Healthcare Services with a Case Study on Cancer Disease in Saudi Arabia
Journal	2022	Xavier 'et al.	Tax evasion identification using open data and artificial intelligence
Journal	2022	Xu et al.	A healthcare-oriented mobile question-and-answering system for smart cities
Journal	2022	Ferreira & Janssen	Shaping the Future of Shared Services Centers: Insights from a Delphi Study About SSC Transformation Towards 2030
Journal	2022	Deng et al.	Big data-driven intelligent governance of college students' physical health: System and strategy

4   Discussion

4.1   For the Efficiency of E-Government Services by the Artificial Intelligence and Machine/Deep Learning

Artificial intelligence and machine/deep learning can play a significant role in improving the efficiency of e-government services in several ways:

Chatbots: AI-powered chatbots can automate routine tasks such as answering frequently asked questions, providing information about government services, and even assisting with simple transactions. This can reduce wait times for citizens and free up government employees to focus on more complex tasks [28, 29].

Predictive Analytics: AI algorithms can be trained to predict which services are in high demand and allocate resources, accordingly, reducing wait times and improving the overall user experience [25–27].

Fraud Detection: MDL models can be used to detect fraudulent activities in real time, ensuring the integrity of government services and protecting citizens' personal information.

Natural Language Processing: NLP can be used to extract information from unstructured data sources, such as text and voice, making it easier for government agencies to access and analyze large amounts of data.

Personalization: AI can be used to provide personalized recommendations and guidance to citizens, based on their individual needs and preferences.

Process Automation: AI can be used to automate routine tasks, such as data entry and document processing, reducing the workload of government employees and freeing up their time to focus on more complex tasks.

Table 4. Using artificial intelligence & machine/deep learning for e-government services

Research	Potential Use of AI&MDL in E-Government Services
Study 1	Using MDL to automate document classification and routing, improving response time and accuracy of service delivery
Study 2	Using AI to predict citizen service demand and optimize resource allocation for service delivery
Study 3	Using MDL to analyze citizen feedback and sentiment for service quality and identify areas for improvement
Study 4	Using AI to automate chatbots for citizen support and service delivery
Study 5	Using MDL to analyze service data and identify patterns for predictive maintenance of infrastructure and services

Table 4 is summarized the potential use of artificial intelligence and machine/deep learning in improving e-government services. Artificial intelligence and machine/deep learning can be used to automate routine tasks, optimize resource allocation, improve service quality, and even predict and prevent service disruptions. By leveraging these technologies, e-government services can become more efficient, cost-effective, and citizen-centric. The use of AI and machine/deep learning has the potential to greatly improve the efficiency, accuracy, and overall user experience of e-government services.

4.2 The Potential Benefits and Challenges Associated with the Implementation of AI and Machine/Deep Learning in e-Government Services

Based on the reviewed resources, it can be concluded that the implementation of AI and machine/deep learning in e-government services can bring several potential benefits, including:

Increased efficiency: The automation of routine tasks using AI&MDL can significantly increase the efficiency of e-government services, reducing wait times and freeing up employees to focus on more complex tasks.

Improved decision-making: AI algorithms can process and analyze large amounts of data, providing government agencies with valuable insights that can inform decision-making and improve outcomes.

Enhanced security: AI can be used to detect and prevent fraud, ensuring the security of citizens' personal information and the integrity of government services.

Increased accessibility: AI-powered chatbots and personalized recommendations can make government services more accessible to citizens, particularly those with disabilities or limited language proficiency.

Better customer experience: The use of AI to provide personalized recommendations and automate routine tasks can greatly improve the overall user experience of e-government services.

Based on the reviewed resources, it can be concluded that the implementation of AI and machine/deep learning in e-government services can potentially pose several challenges, including:

Enhanced data privacy and security: The processing and storage of large amounts of personal information through AI systems raises concerns about data privacy and security.

Reducing bias and promoting fairness: AI algorithms can perpetuate existing biases and discrimination if not properly designed and trained. This can result in unequal treatment and outcomes for certain groups of citizens.

However, there may also be factors that pose risks related to the AI&MDL:

Implementation Costs: The implementation of AI systems in government services can be expensive and may require significant investment in technology, infrastructure, and personnel.

Lack of Technical Expertise: The implementation of AI in government services requires specialized technical expertise, which may be difficult to find and retain.

Resistance to Change: The implementation of AI in government services may face resistance from employees who fear job loss or are uncomfortable with the technology.

While the implementation of AI&MDL has the potential to greatly improve the efficiency and effectiveness of e-government services, it also presents several challenges that must be carefully considered and addressed.

The Table 5 shows the potential benefits and challenges to implement AI in e-government. The potential benefits of implementing AI and machine/deep learning in e-government services are significant, including improved efficiency, cost savings, and enhanced accessibility and personalization of services. However, there are also several challenges associated with this implementation, including concerns about privacy and security of personal data, resistance to change, and the need for continuous updating and maintenance. Additionally, there are concerns about the potential biases and discrimination in AI algorithms and the need for skilled data analysts and experts to interpret the results.

Table 5. Potential benefits and challenges to implement AI in e-government e-services

Research	Potential Benefits	Challenges
Research 1	Improved efficiency and accuracy of e-government services using MDL algorithms	Concerns about privacy and security of personal data used for training and application of AI systems
Research 2	Reduction in service delivery time, cost, and errors through the automation of routine tasks and decision-making processes using AI	Resistance to change and lack of understanding of AI technology by government employees and citizens
Research 3	Enhanced accessibility and personalization of e-government services using chatbots and natural language processing	Challenges in ensuring the quality and accuracy of chatbot responses, as well as the need for continuous updating and maintenance
Research 4	Improved decision-making and policy formulation using predictive analytics and big data analysis	Concerns about the potential biases and discrimination in AI algorithms, as well as the need for skilled data analysts and experts to interpret the results

4.3 Strategies to Ensure the Security and Privacy of Data Collection Through AI and Machine/Deep Learning in e-Government Services

To ensure the security and privacy of data collected through AI and machine/deep learning in e-government services, several strategies can be employed:

Encryption: Sensitive information, such as personal data, should be encrypted both in storage and in transit to prevent unauthorized access.

Access Controls: Strict access controls should be in place to ensure that only authorized personnel have access to sensitive information.

Data Minimization: Government agencies should collect only the data that is necessary to provide the requested services and minimize the amount of data collected and stored.

Regular Data Audits: Regular data audits should be conducted to detect and prevent any unauthorized access to sensitive information.

Privacy Policies: Governments should establish clear and transparent privacy policies that outline how data will be collected, stored, and used. These policies should be made easily accessible to citizens.

Data Anonymization: Whenever possible, sensitive information should be anonymized to reduce the risk of data breaches and protect citizens' privacy.

Continuous Monitoring: AI systems should be continuously monitored to detect any unusual or suspicious activity that may indicate a security breach.

Employee Training: All employees should receive training on data privacy and security best practices, as well as their responsibilities in protecting sensitive information.

Third-Party Management: Governments should carefully vet and manage any third-party contractors or service providers that have access to sensitive information.

Implementing strong security and privacy measures is critical to ensure the responsible use of data collected through AI&MDL in e-government services. Table 6 is summarized the strategies that can be employed to ensure the security and privacy of data collected through AI and machine/deep learning in e-government services, based on the research mentioned:

Table 6. The strategies to ensure the security and privacy of data collection through AI and machine/deep learning in e-government services

Research Findings	Strategies
Lack of security measures in AI-based e-government services	- Implement robust security protocols and encryption techniques - Establish strict access controls to sensitive data - Regularly conduct vulnerability assessments and penetration testing
Risks associated with using third-party AI solutions	- Conduct thorough risk assessments before selecting third-party AI solutions - Establish clear data-sharing agreements with third-party providers - Regularly monitor and audit third-party solutions
Concerns regarding the transparency and interpretability of AI systems	- Develop explainable AI systems that provide clear explanations for their decision-making processes - Establish clear guidelines for the use of AI systems in e-government services - Regularly review and audit AI systems to ensure their transparency and accountability
Risk of bias and discrimination in AI systems	- Establish diverse and representative teams to develop and implement AI systems - Regularly audit AI systems to detect and mitigate any bias or discrimination - Establish clear guidelines for the use of AI systems in decision-making processes
Need for data protection and privacy measures	- Implement strong data protection measures, including data anonymization and pseudonymization - Ensure compliance with relevant data protection laws and regulations - Establish clear data sharing agreements with stakeholders and users - Provide transparent privacy policies to users

It's important to note that these strategies are not exhaustive and may need to be tailored to the specific context and requirements of each e-government service. Additionally, these strategies should be implemented in conjunction with broader security and privacy measures, including regular training for staff and stakeholders, and proactive engagement with users and civil society groups to build trust and ensure accountability.

5   Conclusion

AI&MDL has the potential to significantly enhance the efficiency and effectiveness of e-government services. E-government services refer to the delivery of government services through electronic means, and the integration of AI&MDL can help to provide more personalized and efficient services and improve the overall experience for citizens. In this study, we explore the potential of AI&MDL for enhancing e-government services through a comprehensive review of the available literature. The study focuses on three key research questions: (1) How can AI&MDL be used in the efficiency of e-government services? (2) What are the potential benefits and challenges associated with the implementation of AI&MDL in e-government services? (3) What strategies can be employed to ensure the security and privacy of data collected through AI&MDL in e-government services? The review of the literature reveals that AI&MDL have the potential to significantly improve the efficiency of e-government services by reducing wait times, automating manual processes, and providing citizens with more personalized services. AI&MDL can also help for the accuracy and speed of decision-making, and reduce the costs associated with manual processes. The implementation of AI&MDL in e-government services is also associated with several benefits, including improved service quality, increased transparency, and enhanced citizen engagement. However, the implementation of AI&MDL in e-government services also presents several challenges, such as the need for significant investments in technology and infrastructure, the risk of security and privacy breaches, and the potential for unintended consequences such as job loss. To mitigate these challenges, it is important for governments to carefully consider the security and privacy of sensitive data, and to implement strategies to ensure that AI&MDL systems are transparent, accountable, and secure. In conclusion, the integration of AI&MDL in e-government services has the potential to significantly enhance the efficiency and effectiveness of government services. However, careful consideration must be given to the security and privacy of sensitive data, as well as the potential challenges associated with the implementation of AI&MDL in government services. Governments and policymakers must work together to ensure that the benefits of AI&MDL are maximized while minimizing any potential risks. This study provides valuable insights into the current state of AI&MDL in e-government services and offers important considerations for governments and policymakers. Some potential areas for further investigation could include:

1. Integration of AI and blockchain technologies in e-government: While there are a few articles that touch on these topics separately, there may be an opportunity to explore the potential benefits and challenges of combining these two technologies in the context of e-government.
2. Ethics and governance of AI in the public sector: While some articles touch on the topic of governance, there may be a need for more research into the ethical considerations surrounding the use of AI in the public sector, as well as how to ensure that AI systems are transparent and accountable.

3. Interdisciplinary approaches to e-government research: Many of the articles in the list appear to be authored by researchers from the fields of computer science and information systems. There may be an opportunity for more interdisciplinary research that draws on expertise from fields such as political science, public administration, and law to inform e-government policy and practice.

References

1. Joiner, I.A.: Chapter 1 - Artificial intelligence: AI is nearby. In: Joiner, I.A. (ed.) Emerging Library Technologies, pp. 1–22. Chandos Publishing (2018). https://doi.org/10.1016/B978-0-08-102253-5.00002-2
2. Kalali, A., Richerson, S., Ouzunova, E., Westphal, R., Miller, B.: Chapter 16—Digital biomarkers in clinical drug development. In: Nomikos, G.G., Feltner, D.E. (eds.) Handbook of Behavioral Neuroscience, vol. 29, pp. 229–238. Elsevier (2019). https://doi.org/10.1016/B978-0-12-803161-2.00016-3
3. Gaur, L., Ujjan, R.M.A., Hussain, M.: The influence of deep learning in detecting cyber attacks on E-government applications [Chapter]. In: Cybersecurity Measures for E-Government Frameworks. IGI Global (2022). https://doi.org/10.4018/978-1-7998-9624-1.ch007
4. Babaoğlu, C., Akman, E., Kulaç, O.: Handbook of research on global challenges for improving public services and government operations (2021). https://services.igi-global.com/resolvedoi/resolve.aspx?doi=10.4018/978-1-7998-4978-0. IGI Global. https://www.igi-global.com/book/handbook-research-global-challenges-improving/www.igi-global.com/book/handbook-research-global-challenges-improving/244637
5. Malodia, S., Dhir, A., Mishra, M., Bhatti, Z.A.: Future of e-government: an integrated conceptual framework. Technol. Forecast. Soc. Chang. **173**, 121102 (2021). https://doi.org/10.1016/j.techfore.2021.121102
6. Lopes, N., Rao, H., McKenna, S., Yang, S., Estevez, E., Nielsen, M.: Pannel: digital transformation impact on society. In: Teran, L., Meier, A., Pincay, J. (eds.) University of Texas System, WOS: 000492024900062, pp. 19–21 (2019)
7. Ferreira, C., Janssen, M.: Shaping the future of shared services centers: insights from a delphi study about SSC transformation towards 2030. J. Knowl. Econ. (2022). https://doi.org/10.1007/s13132-022-01072-0
8. Ju, J., Liu, L., Feng, Y.: Citizen-centered big data analysis-driven governance intelligence framework for smart cities. Telecommun. Policy **42**(10), 881–896 (2018). https://doi.org/10.1016/j.telpol.2018.01.003
9. Dominguez, H., Mowry, J., Perez, E., Kendrick, C., Martin, K., ACM: Privacy and information protection for a new generation of city services. In: SCC 2019, Proceedings of the 2nd ACM/EIGSCC Symposium on Smart Cities and Communities (2019). https://doi.org/10.1145/3357492.3358628
10. Soomro, K., Bhutta, M., Khan, Z., Tahir, M.: Smart city big data analytics: An advanced review. Wiley Interdisc. Rev. Data Min. Knowl. Discov. **9**(5) (2019). https://doi.org/10.1002/widm.1319
11. Zhang, J., Qi, X., Myint, S., Wen, Z.: Deep-learning-empowered 3D reconstruction for Dehazed images in IoT-enhanced smart cities. CMC-Comput. Mater. Continua **68**(2), 2807–2824 (2021). https://doi.org/10.32604/cmc.2021.017410
12. Zekic-Susac, M., Mitrovic, S., Has, A.: Machine learning based system for managing energy efficiency of public sector as an approach towards smart cities. Int. J. Inf. Manag. **58** (2021). https://doi.org/10.1016/j.ijinfomgt.2020.102074

13. Zhang, Y., Geng, P., Sivaparthipan, C., Muthu, B.: Big data and artificial intelligence based early risk warning system of fire hazard for smart cities. Sustain. Energy Technol. Assessments **45** (2021). https://doi.org/10.1016/j.seta.2020.100986

14. Kaginalkar, A., Kumar, S., Gargava, P., Niyogi, D.: Review of urban computing in air quality management as smart city service: an integrated IoT, AI, and cloud technology perspective. Urban Clim. **39** (2021). https://doi.org/10.1016/j.uclim.2021.100972

15. Strielkowski, W., Zenchenko, S., Tarasova, A., Radyukova, Y.: Management of smart and sustainable cities in the post-COVID-19 era: lessons and implications. Sustainability (Switzerland) **14**(12) (2022). Scopus. https://doi.org/10.3390/su14127267

16. Al-Taleb, N., Saqib, N.: Towards a hybrid machine learning model for intelligent cyber threat identification in smart city environments. Appl. Sci. Basel **12**(4) (2022). https://doi.org/10.3390/app12041863

17. Hilal, A., Alfurhood, B., Al-Wesabi, F., Hamza, M., Al Duhayyim, M., Iskandar, H.: Artificial intelligence based sentiment analysis for health crisis management in smart cities. CMC-Comput. Mater. Continua **71**(1), 143–157 (2022). https://doi.org/10.32604/cmc.2022.021502

18. Xu, Y., et al.: A healthcare-oriented mobile question-and-answering system for smart cities. Trans. Emerging Telecommun. Technol. **33**(10) (2022). https://doi.org/10.1002/ett.4012

19. Lathrop, B.: The inadequacies of the cybersecurity information sharing act of 2015 in the age of artificial intelligence. Hastings Law J. **71**(2), 501–533 (2020)

20. Hausken, K.: Cyber resilience in firms, organizations and societies. Internet Things **11** (2020). https://doi.org/10.1016/j.iot.2020.100204

21. Rawindaran, N., Jayal, A., Prakash, E.: Machine learning cybersecurity adoption in small and medium enterprises in developed countries. Computers **10**(11) (2021). https://doi.org/10.3390/computers10110150

22. Alexopoulos, C., Diamantopoulou, V., Lachana, Z., Charalabidis, Y., Androutsopoulou, A., Loutsaris, M.A.: How machine learning is changing e-government. ACM Int. Conf. Proc. Ser., Part F148155, 354–363 (2019). Scopus. https://doi.org/10.1145/3326365.3326412

23. Vuppalapati, J., Kedari, S., Ilapakurti, A., Vuppalapati, C., Kedari, S., Vuppalapati, R.: The development of machine learning infused outpatient prognostic models for tackling impacts of climate change and ensuring delivery of effective population health services. In: Baru, C., et al. (eds.) 2019 IEEE International Conference on Big Data (Big Data), WOS: 000554828702102, pp. 2790–2799 (2019)

24. Loukis, E., Kyriakou, N., Maragoudakis, M.: Using government data and machine learning for predicting firms' vulnerability to economic crisis. In: Viale Pereira, G., et al. (eds.) EGOV 2020. LNCS, vol. 12219, pp. 345–358. Springer, Cham (2020). https://doi.org/10.1007/978-3-030-57599-1_26

25. Cavus, N., Mohammed, Y., Yakubu, M.: An artificial intelligence-based model for prediction of parameters affecting sustainable growth of mobile banking apps. Sustainability **13**(11) (2021). https://doi.org/10.3390/su13116206

26. Bharanidharan, G., Jayalakshmi, S.: Predictive scaling for elastic compute resources on public cloud utilizing deep learning based long short-term memory. Int. J. Adv. Comput. Sci. Appl. **12**(10), 73–81 (2021)

27. Zoltán, S., Jinil, Y.: Taxonomy, use cases, strengths and challenges of chatbots. Informacios Tarsadalom **18**(2), 41–55 (2018). Scopus. https://doi.org/10.22503/inftars.XVIII.2018.2.3

28. Narasiman, S.K., Srinivassababu, T.H., Suhit Raja, S., Babu, R.: IndQuery—an online portal for registering e-complaints integrated with smart Chatbot. In: Lecture Notes on Data Engineering and Communications Technologies, vol. 35, pp. 1286–1294. Springer Science and Business Media Deutschland GmbH, Scopus (2020). https://doi.org/10.1007/978-3-030-32150-5_1. https://www.scopus.com/inward/record.uri?eid=2-s2.0-85083664326&doi=10.1007%2f978-3-030-32150-5_130&partnerID=40&md5=0a5228597f831347a8e16aece77de34e

29. Alanazi, S., Kamruzzaman, M., Alruwaili, M., Alshammari, N., Alqahtani, S., Karime, A.: Measuring and preventing COVID-19 using the SIR model and machine learning in smart health care. J. Healthcare Eng. **2020** (2020). https://doi.org/10.1155/2020/8857346

30. Dreyling, R., Jackson, E., Tammet, T., Labanava, A., Pappel, I.: Social, legal, and techni-cal considerations for machine learning and artificial intelligence systems in government. In: Filipe, J., Smialek, M., Brodsky, A., Hammoudi, S. (eds.) International Conference on Enterprise Information Systems, ICEIS – Proceedings, vol. 1, pp. 701–708. Science and Tech-nology Publications, Lda, Scopus (2021). https://www.scopus.com/inward/record.uri?eid=2-s2.0-85122979555&partnerID=40&md5=50d4baccbd4bae3de8a6b6c8fc74334e

31. Alwabel, A., Zeng, X.: Data-driven modeling of technology acceptance: a machine learning perspective. Exp. Syst. Appl. **185** (2021). https://doi.org/10.1016/j.eswa.2021.115584

32. Kenana, O.B.: Should the governments promote or control development in machine learn-ing and artificial intelligence AI? In: Shelley, M., Akcay, H., Tayfur Ozturk, O. (eds.) Proceedings International Conference on Research in Education and Science, vol. 8, pp. 42–51 (2022). The International Society for Technology Education and Science; Sco-pus. https://www.scopus.com/inward/record.uri?eid=2-s2.0-85145898183&partnerID=40&md5=71417726d1f3b8b80fdb81aacb18c488

33. de Sousa, W., de Melo, E., Bermejo, P., Farias, R., Gomes, A.: How and where is artificial intelligence in the public sector going? A literature review and research agenda. Government Inf. Q. **36**(4) (2019). https://doi.org/10.1016/j.giq.2019.07.004

34. Yfantis, V., Ntalianis, K., Ntalianis, F.: Exploring the implementation of artificial intelligence in the public sector: welcome to the Clerkless public offices. Applications in education. Adv. Eng. Educ. **17**, 76–79. Scopus (2020). https://doi.org/10.37394/232010.2020.17.9

Author Index

A. Kö et al. (Eds.): EGOVIS 2023, LNCS 14149, p. 139, 2023.
https://doi.org/10.1007/978-3-031-39841-4

Printed in the United States
by Baker & Taylor Publisher Services